The Light-Hearted Astronomer

The Light-Hearted Astronomer

by Ken Fulton

AstroMedia
Milwaukee, Wisconsin

Published by AstroMedia Corp.
625 E. St. Paul Ave.
Milwaukee, WI 53202

Printed and bound in the United States of America

Cover photo by Anthony Mikulastik

In the fictitious examples given in this book, any resemblance to individuals, companies, telescopes, or books — living or dead — is accidental.

Library of Congress Cataloging in Publication Data

Fulton, Ken, 1947 –
The light-hearted astronomer

1. Astronomy — Amateurs' manuals. I. Title.
QB63.F86 1984 523 84-6218
ISBN 0-913135-01-1

This (my first book) is dedicated to my lovely wife (Doris) and my two fine sons (Shiloh and Shay) who suffered through the writing process right along with me; to my parents (Lenard and Helen) who bought me my first telescope; to all my crazy friends who stuck by me through thick and thin (mostly thin); to all who paid my wife and me slave wages for mowing their yards so I could just barely afford to pay our bills and put food on the table and still have time to write; to all those nerds who never really believed I could pull this book off (but who now want a free autographed copy); to editor Richard Berry (who took a chance on me); to every dedicated amateur astronomy enthusiast who has kept on keeping on, and survived; and, last but not least, to the Creator of the Universe (Who, as far as I can tell, did an absolutely whiz-bang job).

Contents

Introduction

This book is aimed at two very special types of potential amateur astronomers:

1. Those eager, nervous, and inexperienced tenderfoots who are about to get "ants in their pants" and jump in over their heads physically, financially, and psychologically when they purchase their first telescope.

2. Those not scientifically inclined who feel drawn to aesthetically embrace the celestial wonders, but who tend to shy away because they also feel intimidated by astronomy's mantle of science, or are simply uninterested in that aspect of it altogether.

For the benefit of type one, this book considers the ups and downs of amateur astronomy and their potential effect upon the individual. It is not so much about the soft- and hardware as it is about the person — young or old, male or female — who works his or her tail off to buy that large, portable reflector loaded with all those expensive accessories only to find when it arrives, that the shipping charges were astronomical, he or she can't lift it alone without straining something, has no way to transport it, no observing site to use it, and no place to store it. So what happens? The telescope gathers dust. Spiders weave webs about its spider and admire themselves in the telescope's mirror. The owner broods about all of that money squandered on something he or she can't use. The precision telescope has become only a costly aggravation. The void it left in the bank account worries its owner. Eventually, sick at heart and thoroughly disgusted, the owner sells the telescope for half-price and promises never, *ever* to waste any money on astronomy again. The owner has lost; everyone has lost except the lucky stiff who buys the used telescope.

For the benefit of type two, this book seeks to give encouragement by stressing that there is no prerequisite that dictates you *must* embrace complicated formulae and theories to enjoy observational amateur astronomy. Though it certainly adds depth to the enjoyment, you do not have to understand what you see to be awestruck by the sight. You do not have to be curious about the composition of Saturn's rings to be impressed with their beauty; nor do you have to comprehend the astrophysics involved in stellar formation to thrill at the finely focused images of the four stars which form the famous Trapezium in the Orion Nebula.

Anyone may pursue observational amateur astronomy providing they cultivate a dedication, willingness, and sensitivity to *see.* The opportunities are boundless — but it all begins with the opening of a naked eye. Aesthetic enjoyment for its own sake should not be discounted as a very good reason to pursue observational amateur astronomy; nor should lack of education, funds, or scientific inclination be determining factors.

For both types, amateur astronomy should be first and foremost a visual delight, not a search for answers. More than anything else, it is for the enjoyment of observers, not scientists.

Amateur astronomy is not a bed of roses, however. And that is what this book is all about.

1 Amateur Astronomers — A Strange Breed

I HAVE THIS DEAR OLD CRONY WHO collects leaves and elephantine books. He is seventy-six and living month-to-month solely on his social security income, yet he will brave any terrain to obtain an unusual leaf and let go his precious bucks for just the right tome — the bigger the better in both cases. Believe it or not, the dead leaves and the unread books work hand in hand to make my friend happy. During times of quiet relaxation in the privacy of his modest home, he presses the leaves of exotic and not-so-exotic foliage between the leaves of exotic and not-so-exotic printed matter. It is a harmless hobby, one remarkably free of frustrations and pitfalls. Not once has it ever given my friend any reason to scream, gnaw his fingernails down to the quick, slap himself silly, hate himself for being stupid, consult a shrink, or turn into a babbling vegetable. No one I'm aware of thinks that he is odd for embracing his hobby.

I have this relative who stays up till all hours — most of them ungodly — making very small talk with people all over the globe with the aid of his very expensive, very elaborate, and very impressive state-of-the-art radio communications setup. Usually, he can tell you how much rain they got tomorrow in Penang, Malaysia, or who gave birth to what in Kalgoorlie, Australia. From the beginning, his hobby has been a source of pleasure to him. No one I'm aware of thinks he is odd for embracing his hobby, though some of his neighbors would pay God or the F.C.C. a princely sum to zap his towering antenna into limbo.

I have this very special friend who works like a mule overtime at a lumber yard and spends most of his free time every clear night studying faint smudges of light which are galaxies far, far away. He absolutely hates to waste good observing time even if he's had an exceptionally rough day at work. He's a strong back by day and a tough astronomy enthusiast by night.

His neighbors don't quite know what to think of him. He built this short-walled building in his backyard that has a weird roof. Almost every clear night, that roof somehow rolls aside, and something akin to a giant cannon barrel soon rears its muzzle high. This "cannon barrel" is really three barrels of the 55 gallon variety welded end-to-end. It's as ugly as the south end of a north-bound hippo. I don't think any of my friend's neighbors really believe the thing is a precision telescope. He has, on numerous occasions, invited them over for a look-see. So far, he's had no takers.

They don't quite know how to take him. Late at night, when they get up out of bed to go potty or enjoy a late-night snack, they *know* he's out there in his strange building, fumbling around in the dark, doing God-knows-what. They never see him turn on a light of any kind while he's out there. Sometimes, when there is no Moon, they see a curious red glow that may or may not be just a figment of their imagination.

Shortly after the cannon's appearance on the scene, my friend asked each one to kindly refrain from leaving on their exterior lights after midnight. Because they're good folks, they've been careful to oblige. They don't really understand why he wants it so dark around the neighborhood. He's explained to them that the brightness of the exterior lights would hamper his night vision — but they didn't listen to him. These are the type of people who only check the sky to see if it's going to rain. And they just don't know where my friend is coming from.

You can't blame them. They're not astronomy enthusiasts. They haven't heard the call of the night.

Neither has my friend's wife. She believes he's two cards short of a full deck, though loveably so. He is so obsessed/excited with the prospect of discovering an extragalactic supernova with this homemade 22-inch Dobsonian reflector telescope that, at times, even *I* fear the worst.

His feeling toward me is mutual, I'm afraid. He insists that I'm a trifle out-of-plumb from observing the Moon so much. That's how close we really are. No better friends exist.

Because we both love astronomy and because we've both experienced the ups and downs of astronomy, we've formed this bond, see? We both know first-hand that embracing the hobby of astronomy is a lot like charming a cobra. You do it right, or you get bitten. And sometimes, even if you do it right, you get bitten anyway. The only difference is: *Astronomy's venom will not destroy you unless you let it.* If you desire, you can slowly but surely develop an immunity to the venom.

My friend and I have developed such an immunity — but it took years of blood, sweat, and tears. Now, we are both tough as flank steak. We can shake off the venom and continue with our work. Even so, the memories of our struggling years still haunt us. Staying alive was hard. I made so many stupid mistakes, took so many wrong turns into blind alleys, and suffered so many aggravations that, even now, it amazes me that I still love astronomy.

Why didn't I just chuck it all and take up stamp collecting?

One night when we were both really drunk on starfog, my friend asked me, "What are we *really*?"

Good, deep question, that. One that definitely required some serious discussion. So we retired to his den and guzzled strong coffee and threw words at each other's face for nearly two hours. We'd just about nailed the answer down when, with very little warning, the rising Sun took away the night's magic and forced us to become "normal" people again. If that wonderful night would have just held on for thirty minutes more, I really think the two of us would have figured it all out. We'll always remember that night, and the one answer that got away.

What are astronomy enthusiasts? They certainly have peculiar lusts. Strange things turn them on — dismal points of light, dust-scattered photons, long-distance fluorescence, a lovely rose in Monoceros, gargantuan pinwheels, clusters of fire, darkness at noon, cosmic peekaboo, dilated pupils, tripods, large porthole windows, surplus junk, ray-tracing, a frontal passage, an out-of-order vapor lamp, Teflon, holes that are impossible to see, holes that may not even exist, stellar nurseries, and quiet parties on starlit mountain tops, to name just a few. Astronomy enthusiasts enjoy lurking about in the darkest of dark, vampire-like, mumbling to themselves. Sometimes you'd swear they speak unknown tongues. They come in all shapes and sizes. Some are fat. Some are tall. Some are short. Some are small. Some have bodies like Greek gods or goddesses. Some wear bras; some do not. Some are male. Some, it's hard to tell.

Some are old and famous; some are young and famous. Some want to be famous. Some could care less. Some are highly intelligent, some are middle-of-the-road, and some have a hard time spelling intelligent. Some are good at delving into the secrets of the universe, and some are quite content with simply appreciating the beauty of it all. For some, astronomy is a science; for others it is a soothing balm, or an escape from everyday life, or a tap into the cosmic connection.

And for some, astronomy has been, can be, or will be a horror story the likes of which Poe or Lovecraft or even Stephen King never dreamt.

Astronomy enthusiasts are, to say the least, members of a motley crowd. Red or yellow, black or white, the sky is precious in their sights — that's just about the only thing they all have in common. Some chug-a-lug beer, while others quaff Gatorade, while others sip orange juice, while others enjoy imported wines. Those who combine their love for astronomy with their love for astronautics prefer, of course, Tang. A large slice of the pie almost worships coffee. They believe that there *must* be something magical about the brew. "A good swallow of mountain-grown intensifies feeble light" — that's somebody's law. I'm not gonna tell you whose.

Astronomers, amateur or professional, seldom agree completely on anything. Take any two and both will know just how the universe began, or at the very least have a neat-sounding idea. What's more, both will gladly explain to you just why the other guy's theory is goofy. And both will defend to the death the right of the other to be goofy.

Taken as a group, astronomy enthusiasts exhibit the gamut of human foibles. And because they are human beings, they take much of their humanity with them when they trip the light fantastic, which in this book you may take to mean observe.

Yes, jealousy and envy and covetousness and thievery and deceit have no qualms about attending star parties, dictating the business practices of disreputable telescope dealers or manufacturers, or living together in sin under domes atop high mountains. Astronomy is a field where "Who will be the first to rediscover Halley's comet?" and "Who will have the sexiest telescope at the star party?" and "How can we market a prettier telescope than our competitor's newest model?'' can cause much loss of sleep and a marked increase in the sales of Maalox. At star parties, small but intense wars can erupt without warning, ignited by something so trivial as the old refractor versus reflector controversy. For those who frequently attend such parties, keeping up with the Joneses can become a way of life. This so, they usually attend armed to the teeth where telescopes and accessories are concerned. Some know how to control their weapons and some do not. A novice must be very careful, tread softly, and stay out of all possible lines of fire, lest he or she get riddled in more ways than one. And where the dealers and manufacturers are concerned, let the buyer beware! Competition in the industry and in sales is as keen, if not keener, than in the television networks' ratings wars. Not everyone fights fair, and it is often the innocent, unwary bystander who gets injured by the shrapnel.

To make use of an over-used cliche, it's a jungle out there. Astronomy is no field for a carefree joy ride, or a halfhearted trophy hunt, or a casual starlit picnic. This jungle has teeth, and they can chew you up and spit you out. Unless you allow someone who knows the jungle to help you gird your loins, teach you survival tactics, and turn you into something of an astronomical Indiana Jones, there is a good chance that you are going to rush in where angels fear to tread, find yourself lost, at bay, and at the mercy of all manner of teethy beasties.

Very early on, preferably at the very beginning when you take your first baby steps into the jungle, you will need a book of survival techniques — one that warns you about the many dangers that hide in this special jungle, chameleon-like, and are always prepared to strike you down. To survive, you must cultivate a sense of humor when it comes to astronomy (no easy task), dispense with taking *any* part of astronomy too seriously, be willing to get into shape physically and mentally, be courageous and determined to succeed, and be patient enough to earn your stripes one step at a time.

Consider this book your survival manual. If you take it to heart and happen to be *worthy* (I'll explain this later), you will survive, albeit not without a few scars to brag about during cloudy-night bull sessions.

So what the deuce makes me so qualified to help you out?

I'm a successful astronomy enthusiast right down to my tattered jockey shorts. Find me at a bull session some time and I'll show you my scars and stripes. Give me half a chance and I'll tell you about each one. I'm anything but shy, and modesty is not one of my favorite words. After nearly twenty-five years struggling and surviving and observing in astronomy's jungle, I feel I've earned the right to write this book. At the risk of sounding arrogant, let me say that I am tough and confident now. Me and the universe, we got this bond, see? I'm proud of that, and you will be just as proud when and if you make the grade.

The folks who published this book and I sincerely want you to make the grade. Since you are reading this book, chances are you are toying with the idea of becoming an astronomy enthusiast, or have been stung one way or another during your first efforts to embrace the hobby.

First off, let me urge you to get comfy. Loosen your tie. Better yet, take the silly thing off and drape it somewhere out of sight. Now, take your shoes off. Nobody's looking. And if you are female, do whatever females do to loosen up, within reason.

I want you to relax with this book.

And too, I want you to laugh while reading this book. If — God forbid! — you are devoid of a sense of humor and feel that the distinguished science of astronomy has no place for rib-tickling in its hallowed halls, you'd better stop reading right here and check out someone's doctoral thesis instead. This is not going to be your average, dry, passionless dissertation chock-full of technical pudding that will go bad two days before publication.

If you do have a sense of humor, and if you want to trip the light fantastic and have an easier time doing it, keep turning these pages.

It is very important to me that you survive as an astronomy enthusiast.

Why?

A lot of reasons, but one in particular:

If everyone on this planet would really get involved in astronomy, we would not have to worry about World War III. A nuclear holocaust would make for very bad "seeing."

2 Astronomy's Jungle

THROUGHOUT THIS BOOK, I AM going to take the liberty of putting words into your mouth. If you find you can't stand the taste of them without barfing, feel free to spit them out — but try your best to roll with the flow, okay? It's for your own good, believe me.

"Should I, or shouldn't I, pursue astronomy?" you ask.

Or course you should! For the life of me, I can't imagine why anyone would not at least give it the old college try.

"But you said it was a jungle."

Ah, but *what* a jungle! One full of galactic adventures, quasi-stellar intrigues, nebulous victories, cosmic discoveries, and astronomical challenges the likes of which Tarzan never dreamed. Certainly, there are some risks involved, but that's true of eating a can of tuna. You will be required to invest quite a bit of yourself in this boundless enterprise. And if you invest wisely, you will reap enormous profits. I'm not talking money here. I'm talking a sense of accomplishment, hobnobbing with immensity, tapping into tranquillity, and tasting the infinite as it rolls like priceless wine on the tongue. I'm talking profits that settle deep in the heart and accrue interest with each passing moment.

"Then darn the torpedoes, full speed ahead!"

That's what I like to hear. Good for you! But let's not forge ahead at full speed just yet. Slow and easy is right for now. Remember, you are a tenderfoot.

"So how do I begin?"

Imagine that you are standing at the edge of astronomy's jungle. All looks quiet and peaceful, although somewhat formidable. You listen to hear the growl of the teethy beasties, but you do not hear them. They prowl a bit deeper in the jungle. Believe me, they're there.

You see two paths leading in. You must choose one path and enter.

"Why two paths?"

Because this is *my* book. I'm free to set the scene here. Just trust me.

"But what if I don't enjoy the path I choose?"

No problem. At many places in the jungle, the two paths cross. You can switch over if you've a mind to, or make your own if you've got the bravado.

"So how can I make such a decision this early in the game?"

Good question. Notice that one path is wider than the other. The wider path is the Aesthetic Path; the narrow one is the Scientific Path. Before you choose, you must do some serious soul-searching. Take a long, hard, honest look at yourself. Leave your rose-colored glasses in your backpack.

Ask yourself questions like the following:

Are you patient and open-minded? Do you usually finish what you start? Do you keep on keeping on in spite of obstacles that block your way? Are you a fighter who gets up quickly when knocked down? Are you better than average at math? Do you regularly read and enjoy science-related books and articles? Are you genuinely interested in the structure and origin of the universe? Do you or *did* you make better than average grades in your science classes in school? Are you interested in knowing why the night sky is so beautiful? If you could spend two hours on one of Jupiter's moons, would you want to spend time conducting scientific experiments? If the universe was a frog, would you want to dissect it and analyze it?

If you answered "yes" to the questions above, I think it is safe to say that you are scientifically inclined. That path should not prove too traumatic for you.

On the other hand, if you answered "no" to most of them, give the following questions a go:

Do you have the sensitivity of a poet? Have you ever felt the desire to write a poem about the universe? If you could spend two hours on one of Jupiter's moons, would you prefer to be left alone with your thoughts. Are you gravitating to astronomy because you feel tripping the light fantastic just might be therapeutic for you, a good way to escape the hassle of the day? Does merely hearing the words "differential calculus" give you a mental block and a severe case of the mathematical shakes? In school, were science courses your most difficult subjects not because you were uninterested, but because you simply could not tackle the technical terms, theories, and mandatory computations? If the universe was a frog, would you not give a farthing to know what makes it croak and jump, but give your last shekel to merely enjoy hearing it croak and watching it jump? Do you have other hobbies that already take up a great deal of your time? Do you need at least six hours of sleep a night to function the next day? Are you a hopeless romantic seeking romance far from the test tube?

Did you answer "yes" to most of these questions? If you did, I suggest you take the Aesthetic Path. To be sure, it is the safest way, littered with the least number of obstacles. It will steer you clear of the jungle's densest terrain and guide you through those portions with the most relaxing scenery.

Simply read the handwriting on the wall. You know yourself better than anyone. Weigh the evidence and choose your path.

"Let me get this straight. You're telling me that I don't have to be Joe Genius, or Joe Scientist, or Joe Honor Student to become an astronomy enthusiast?"

Correct. It is not necessary to know how to spell pirouette in order to enjoy a ballet. Ergo, it is not necessary to embrace the science in astronomy to enjoy astronomy. You do not have to be an expert in the interstellar medium or even know what it is to appreciate the beauty of the Trifid Nebula. Knowing about it adds icing to the cake, but it is not mandatory. Likewise, you do not have to be a whiz at math to feel a surge of excitement the very first time you observe the Moon telescopically. You do not have to understand Moon morphology to thrill at all that "magnificent desolation," as astronaut Buzz Aldrin so aptly described it. What's more, you don't have to study orbital mechanics to discover a new comet.

Let's face it. A lot of you out there are whiz kids — but a whole pack of you, through no fault of your own, are just not scientifically inclined. Some of you would never attempt to understand the chemical composition of Jupiter's atmosphere because, quite frankly, just trying to pronounce the additives forced into the average potato chip makes your brain reel. Does this mean that you are unqualified or unworthy to observe Jupiter's everchanging cloud bands or pig-out on a bag of chips? Of course not! It simply means that you must use your head when trekking astronomy's jungle and make sure you don't go in over it.

"What if I didn't, uh, corner the market on brains? This is just a hypothetical question, you understand."

I understand.

It makes no difference to the universe if you are Joe Genius, Joe Average, or Joe Dumb. It also makes no difference to her if you are Joe Minority, Joe Rich, Joe Poor, Joe Unlucky, or Joe Female. If you were blessed with eyes that can see, you are free to try to become an astronomy enthusiast. There are no entrance exams, no blood tests. There is no one who can tell you that you do not qualify — except yourself. You simply must have the burning desire to survive despite the obstacles, the sensitivity to appreciate the majesty of it all, and the willingness to accept the no-pain, no-gain rule of thumb. Not all individuals can become astronomers — but anyone who really desires to can become an astronomy enthusiast. Psalm 19:1 states: "The heavens declare the glory of God; and the firmament sheweth his handiwork." That declaration is not selective, and the show is for everyone who will see.

Search your soul, friend. Then, choose your path.

"Did you choose the right path at first?"

No. Differential calculus gave me hemorrhoids. The final was too much of a strain on me. It was either give up my dream of becoming a Nobel Prize-winning astrophysicist or spend all the prize money on Preparation H.

I saw the writing on the wall early and turned to something that came easier for me: writing. Instead of giving up and walking away from astronomy, I approached it from another direction. I would not have survived very long wandering the scientific path in the astrophysics section of the jungle.

Even so, I rambled alone and very frightened for a long while. There were many times I wanted to run from the jungle, but my respect for the universe kept me from it. I kept on keeping on until I found my special section.

Wanna give it a try? Then take my hand. Here we go.

3 First Things First

CONGRATULATIONS. YOU ARE NOW in the jungle.

No matter which initial path you chose — aesthetic or scientific — notice that your path branches out ahead into quite a few routes of decision. The chances are you will gravitate to either of those two well-lit pathways right in front of your nose. Both look harmless enough, don't they? One has the potential of scarring you for life or turning you completely against astronomy; the other can either usher you into a marvelous new life style, or scramble your brains and fricassee your bank account very quickly.

The pathway most likely to destroy your interest is the much-traveled route to the department store, a business designed to take control of your senses the moment you enter. It is here, in the company of slicers, dicers, choppers, and *Star Wars* action figures that an object can be had which, every clear night, causes Galileo to spin clockwise in this grave. Using the cleverest of nomenclatures, the family of astronomers has termed this object "The Department Store Telescope." Translated into layman's terms, that means junk.

Imported 1000-power refractors with objective lenses of factory-reject gelatin are most common. A lot of these useless things can be seen embarrassing Christmas trees during Yuletide, used as nifty props in wretched made-for-television movies, and as furniture adorning the modern-day living rooms pictured on the covers of interior design magazines. That's just about all they're good for, unless you count using the tube as a whiffle-ball bat. They may appear attractive — but so do coral snakes. You'd think they would be much less expensive than a much better quality telescope of identical size, but in reality many cost more than higher quality, non-department store scopes twice the size.

Take my advice and steer clear of these beasties like the plague. Ignore their siren calls. Stuff cotton in your ears. Buy a blender two aisles down. At least, you can do something with a blender.

"If these telescopes are so bad, how come they sell?"

Because they're there within easy reach. Because they haven't had any quality competition; the average customer has not been able to shop and compare. Because the cardboard boxes these telescopes usually come in are plastered with extravagant claims, most of which are outright lies and an insult to truth in advertising. Because these optical travesties are usually purchased by uninformed, non-descriminating buyers.

I'm happy to report that, as of this writing, things are looking up. Better quality telescopes are beginning to find their way into the better camera stores of the major cities, thanks to the continued growth and outreach of some of the larger, more quality-minded companies. However, unless you know which are the more quality-minded companies, you could still get stung.

I suggest you choose the path I'm about to tell you about. This path will introduce you to the real world of astronomical hardware. You will come to it while trekking either the Aesthetic Path or the Scientific Path.

Right off, you should encounter a magazine rack of sorts. If not, search until you find it. Your public library should be able to guide you. You will know it when you see it. This rack, though small, will be stocked with magazines specializing in the field — excellent ones like ASTRONOMY, *Deep Sky*, *Odyssey* (for ages 8-14), *Telescope Making*, and the venerable (but somewhat advanced) *Sky & Telescope*. These excellent publications are not only jam-packed with interesting, inspirational, and educational articles (some heavy on the science, some not), observing aids, monthly star charts, helpful hints, news items, and astrophotos that will knock your eyes out, but they are also spiced generously with ads — classified and display — hawking all manner of goodies.

You will probably go bonkers reading the ads. You will salivate while looking at all the merchandise pictured. Your heart will pound. Before you know it, you will be reaching for your charge card or checkbook.

Stop right there! Freeze!

Let me prepare you first. There are some things you need to be aware of concerning the ways telescope and accessory manufacturers and dealers advertise their wares.

For example, you should know that most companies would have you believe that they sell the best telescope or accessory available *in its price range*. My, oh my, the manufacturers work those four little words to death in the ads and catalogs. I've encountered them so often that I just refuse to even see them anymore, no matter how bold the print. How on earth is that silly, useless claim going to help you choose a scope when the prices from company to company are intentionally staggered? Just try to find two competitive telescopes of identical size priced and equipped identically. Also, in many cases, the item is promised to be as good or better than competitive items selling at *twice* the price. You should wonder, then, why anyone in their right mind would do business with the higher-priced company. I mean, why pay twice the price for something when you can get a like product elsewhere for half the outlay? And it will doubly confuse you to read that the company with the higher price has been in continuous business for forty-six years. Must be a lot of fools around throwing their money away . . . right? Wrong. The fools are throwing their money away on those department store models, remember?

As you thumb breathlessly through the pages of the magazines, take the ads with a grain of salt. They're not plastered with lies, mind you. They are ripe with clever phrases, ingenious come-ons, fine-print disclaimers, technical mumbo jumbo, fantastic astrophotos (usually taken by an expert with the telescope or telescopes pictured), impressive layout, and subliminal and not-so-subliminal challenges and rebuttals aimed directly at the jugulars of their competitors. Oddly enough, some of the most interesting reading is buried within the ads.

You should know the companies are "fishing." When you go fishing, you take along the best bait available, don't you? Well, when the manufacturers and dealers prepare their ads, they're fishing for you and your money. They choose their bait very carefully. You should do the same before you bite. Once you take the bait, the hook is set from the other end. Don't blame them for reeling you in. And don't blame them for taking advantage of the best bait.

Let's study the bait a bit, okay?

4 The Ads

THE MAGAZINE ADS COME IN FOUR sizes: small, medium, large, and X-large. Just like T-shirts. (You're impressed with that little tidbit of info, aren't you? Bet you're wondering who does all my research, huh? Bet I could knock you over with a feather, couldn't I?) And just like T-shirts, some are plain, some are tastefully decorated, and some will make you want to heave your cookies — but all of them cost money.

The manufacturers and dealers spend all that money to get noticed. So let's notice them for a while, okay? And let's leave the big ones till last.

The Small Ads

Consider the one- or two-man operations. In most cases, their limited funds force them to purchase small ad space. In a few cases, the small operation is so successful that it really does not have to advertise, yet does so in this modest manner only as a courtesy to those who, somehow, have escaped the word-of-mouth stuff.

Obviously, there's not a whole lot of room to toot horns in these ads. There is enough space to briefly state the nature of the service rendered, the firm's name, address, and telephone number. If there is room, many will squeeze in something impressive like "In Business for Over Forty Years" or "Color Brochure Available" or "Send SASE for Mimeographed List (Please Add $4.75 to Help Cover Mailing Costs)" or something catchy like "Optician for the Stars" or "Member O.A.A. (Obese Astronomers of America)" or "Prompt Delivery" (whatever that means) or "Biggest Little Telescope Manufacturer in the Free World" (whatever that means).

Usually, these small advertisers specialize, or concentrate their expertise, in certain areas. Like rich-field telescopes, for instance. Or eyepieces. Or heated underwear. Or inflatable domes (decorated with the Disney character of your choice — no extra charge). Or focusers ("We're talking *focusers* here"). Or finders (which can cost you more than your telescope). Or elliptical optical flats (flatter than flat). Or special mirror coatings (103.6877546% reflectivity). Or books (hopefully hawking this one). Or esoteric bric-a-brac (no comment).

The small ads can be gold mines. The problem with them is this: you can't really get a feeling about the grade of the ore from reading the ads. I mean, are they really experts at what they do, or are they just a couple of dudes trying to make a buck on the side? Or worse, is it just one expert in mail fraud having one more profitable fling before he retires to Argentina?

"Yeah! Just how safe *am* I patronizing these small outlets?"

Very safe, if they've been in business for a good while. Believe me, rotten apples don't stay around long in astronomy's kitchen. If the apple has been advertising regularly for three years, you're on pretty safe ground. Ten, it's fairly obvious they're delivering the goods. Thirty, you'll be lucky to get your order in six months. Firms with this length of tenure usually cannot supply the demand immediately.

"So how am I to know just how long they've been advertising?"

Simple. Write the magazine carrying the ad and ask them — preferably long before you shell out any money. As a courtesy, enclose enough postage to cover their reply. Believe me, if they smell a bad apple, they will let you know. (Or look up a few back issues of the magazines in your public library. Notice how many outfits advertising three years ago aren't advertising now? *Caveat emptor*.) The magazines are very interested in protecting their readers from unfair business practices, especially when it concerns one of their clients. They are willing to help, but *don't* expect them to endorse any advertiser. Ideally, the fact that they carry the ad at all should be enough endorsement — but do not underestimate astronomy's Dark Force. Rarely, it turns good apples into bad apples, quite out of any mere magazine's control.

Along this line, let me stress that the magazines are not in the position to mediate unfortunate disputes, but they can and will refuse to continue running an ad for any outlet that does not deliver the goods as advertised. Be advised that, in the magazine business, doing so is a drastic measure which must be taken with extreme caution. They must take the time to gather sufficient hard evidence to protect themselves and to legally justify such an action. In the meantime, x number of unlucky customers must pay the price.

You can protect yourself by not patronizing any outlet until you've checked them out. Do not patronize any *new* outlet until it has had sufficient time to prove its integrity. I know this sounds cruel, especially to a new business, but hey, nothing I say is going to completely stop people from buying a sexy new product in this crazy day and age, *especially* at a low, introductory price.

"If the long-lived outlets are really good, why are they still small?"

Easy answer: They like it that way! Most have graduated with honors from the old school of thought which taught that small businesses are good businesses. They provide an honest living *sans* the headaches and office politics that accompany the mass marketing of a precision product. They provide a work atmosphere conducive to craftsmanship and easy breathing.

Be advised, also, that such an atmosphere can also result in something similar to the following:

> "Yes, Mr. Beginning Astronomer, we can have your perfect mirror ready in ten years, plus or minus a year."
>
> "That long, huh? How about one just a shade short of perfect?"
>
> "Sorry, but we do not make them that way."
>
> "What if I offered you double to get me one fast?"
>
> "Sir, surely you are joking! We *are* an honorable firm."

Make no mistake about it, there are still artisans out there who are content with staying small, businesswise. They are gifted craftsmen who take pride in their work and deliver an excellent product. They could set their own salaries working for the big guys, but they choose not to. And not one of them is starving for business. In many cases, you will have to take a number and wait your turn. In many cases, the prices they charge are higher than for similar items produced by the big guys. Yet, by and large, they are worth the extra money. I will pay more for a custom, hand-figured, artisan-crafted mirror any day. Perhaps I'm a hopeless romantic, but a lot of pride and personal hands-on attention make for a better mirror in my book. And this *is* my book.

So don't let the size of a small ad turn you away. Just do your homework first. Play detective before you buy. Case the joint first.

And if a new ad should try to lure money from your purse, resist the temptation for a while. There are plenty of antsy people out there who simply must have everything new when it concerns astronomical equipment. Let them take the risks. If the new firm delivers a great product, rest assured that they'll be around for a long while. If not, word will spread like wildfire, and they'll turn to ashes quickly. There's no sense in putting your head on the chopping block when others are willing to do it for you.

The Medium-Sized Ads

These ads are the ¼- to ½-page dudes. They aren't cheap, especially in the slick magazines with large circulations. If an outlet has sunk this kind of money into an ad campaign for six months or so, you can be fairly sure the business is not something someone is running in the back of his garage to supplement his income from a newspaper route. In the main, these guys are dead-set on becoming bigger.

In such ads, you will come across claims like "Fastest Growing Telescope & Accessory Maker on the Banks of the Mississippi" or "Central Texas' Optical Rodeo" or "Today Iowa, Tomorrow the World!" or "Big Business with Personal Touch" or "Lay-Away Plan!" or "Virtually Off-the-Shelf Delivery" or "Free T-Shirt with Every Scope Purchased (22-inch or Larger)."

Uniquely, these ads reflect the character(s) of the owner(s). Study them and you will see this is true. Some are graphically pleasing; some are graphically embarrassing; some will remind you of a grocery ad. Some use photos: a mug shot of the outlet's owner, or the owner's baby daughter, or the owner's girl friend, or a glamour pic of the outlet's most attractive product.

Generally, these ads will give you the feel of shopping at a local home-owned and -operated business. There are quite a number of items to choose from, yet the atmosphere is still up-close and personal. They are often authorized dealers for the big guys, and their prices usually reflect a keen competition.

If you're a bargain hunter, you should stalk these outlets. They'll talk turkey with you, especially if they're hungry. If you will take the time to bone up on your turkey talk before you talk it, you will probably walk away happy and with your game bagged.

Again, do your homework.

The Large and X-Large Ads

Here is where all the stops are pulled out, if you will forgive the technical pun. We're talking big, big business here, folks. We're talking about the marketing of observing packages that have cost the companies unbelievable bucks just to research. Where the telescopes are concerned, we are not talking the equivalent of Rolls Royces, but we *are* talking Oldsmobiles, Buicks, and Cadillacs. Perhaps even Mercedes-Benzes.

There are giants in the amateur astronomy industry — ones who work very hard to set the trends. No one in his or her right mind would *not* agree that these major outlets make their telescopes and accessories affordable, portable, and sexy. State-of-the-art optics, however, is not their bag. No matter what their ads claim, they're selling burgers and fries, not filet mignon and potatoes au gratin. Like Burger King, what they serve is palatable, but not *haute,* cuisine. The selection of dishes served will suit almost every appetite. The service is undeniably fast.

And when they run ads, they run *ads.*

It was different in the 60s when I got hooked on astronomy. I guess I would label yesteryear's advertising practices sedate. Then, the major outlets usually stuck with the same ad year after year. Only the prices changed with any degree of regularity. The birth of a new telescope or an improved model was a big, big event. Still, there were no fireworks or marching bands to herald the new arrival — only a calm, composed, and dignified announcement. You had the feeling, then, of being in a classy auction gallery, and a nod of approval was the proper reaction.

Now, it is another ball game entirely. Today's full-page, double-page, and even triple-page ads scream at you. They are seldom the same from month to month, it seems. The competition for your patronage is now so keen that no major outlet can afford *not* to stir the waters regularly to attract attention. To survive, they must commit themselves to running a wild and wooly, nerve-wracking race to come up with something, *anything,* new or exotic or, when all else fails, prettier.

They've invested BIG bucks. They've built BIG plants. They've hired LOTS of employees. Consequently, they must SELL a lot of telescopes to stay in the black. They not only need new customers, but they must have repeat customers. This so, they would rather you *not* stay satisfied with your new telescope (even if it is theirs). They want you to do exactly what auto manufacturers want you to do: buy a new, improved model every two years or so. They may guarantee their telescopes for ten years, twenty years, or fifty years — but they sure as shootin' do not intend to let you stay happy with yours for that long, no matter who made it.

It makes no difference if you are a potential new customer or a potential repeat customer, they will do their best to seduce you with their ads. If that doesn't work, they will try to bombard you into compliance with extravagant claims. If that doesn't work, they will try to confuse you into submission with almost unintelligible technical jargon. As a last resort, some of the more ruthless outlets will use their entire ad space to take cheap shots aimed at damaging their closest competitor's credibility. The resulting point-counterpoint can make for really fun reading — even if you don't know what in Pluto's name they're arguing about.

Here is a slightly (but only slightly) exaggerated example of what I mean:

Fictional Precision-Eye Optics has just come out with a new model scope of the Fulton-Horton design (don't waste your time looking for this design in any technical journals). So has Betelgeuse Optics. (Ever heard of industrial spies?)

Both firms are giants in the business. They are arch-competitors. When Precision-Eye publishes a full-color 64-page catalog, good old Betelgeuse rushes to press with a 65-pager and throws in a poster of Mount Palomar Observatory in heat.

Miraculously, both full-page ads for the Fulton-Horton design come out in the same issue of your favorite monthly astronomy magazine. Presume that you ordered both catalogs and now have them in hand. Since good old Betelgeuse gave you a free poster, you read their catalog first. I mean, fair is fair, right?

Immediately, you get the feeling that you need to decide what you want very quickly since the prices are listed on a separate sheet of biodegradable paper that has a warning in bold print stating: **Introductory Prices Subject to Change without Notice**.

You check your watch.

Quickly, you compare catalogs and notice that Precision-Eye's 8-inch Fulton-Horton is priced $19.65 cheaper than Betelgeuse's 8-inch of the same design.

Suddenly, the poster is not as impressive. A nice touch, granted, but not $19.65 nice.

Now, you are leaning toward Precision-Eye's model. (Didn't know you were that fickle, did you?)

You study the catalogs even closer.

Aha! Precision-Eye charges $17.00 more for their tripod. And you've got to have a tripod unless you want to observe on your belly all night.

You reach for your Texas Instruments calculator.

$19.65 − $17.00 = $2.65.

Very interesting. That's where good old Betelgeuse is getting their money back for that "free" poster.

So everything so far is even.

Still, you like the color of Betelgeuse's tube better than Precision-Eye's. It'll go better with the color scheme of your storage closet.

So that clenches it. You make up your mind to buy from Betelgeuse.

"I don't understand. What has this got to do with cheap shots in advertising?"

I'm getting there. Trust me.

Well, you've made up your mind. You are relaxed now. You settle back to browse through your favorite astronomy magazine. The most recent issue arrived with the catalogs in today's mail.

Precision-Eye's full-page ad catches your eye. They've added something new. You read the following:

Notice to Our Valued Customers

Recently, one of our competitors advertised an 8-inch telescope of the Fulton-Horton design which they claimed is far superior to any other telescope of the Fulton-Horton design now on the market. Since we are the only other firm marketing such a telescope, we feel it is our duty to defend our fine model. Please be advised that our model's bi-pentafical wertner exercises the flibis while guarding the sympathetic stop from excessive genuflexure. Our competitor's model does not have such a wertner. Also, our larger *American-made* slipsink filbert makes for light grasp equivalent to a telescope an eighth-inch larger!

Holy Erfles! How stupid of you to even consider Betelgeuse's scope. It doesn't have a wertner! And who knows what clowns made the slipsink filbert. And anyway, Precision-Eye's is somehow an 8⅛-inch aperture 8-inch telescope!

Gosh! You almost made a bad mistake.

Relieved now, you decide that Precision-Eye's scope is the one for you. Part of you wants to order, but your gut tells you to hold off a while.

Three months later, you notice Betelgeuse Optics' full-page reprisal. It reads:

Note to Our Valued Customers

The inclusion of any type of wertner, and especially a bi-pentafical wertner, causes undue stress on the flibis which, in turn, tends to encourage photon splatter (see illustrations of ray tracings). Also, our German-made slipsink filberts have been precision-machined on American-made lathes to the highest tolerances and must be able to pass the highly respected Urbantke-Woolsey test with a grade of A- or better before they are allowed into our manufacturing facilities. Detailed results of this test are available to any interested who can read German.

Ah! You never even thought of photon splatter, did you? Admit it.

Now, you put down the magazine and think, "I don't know what on earth photon splatter is! Or what a slipsink filbert is! And who in the hell are Urbantke and Woolsey?"

Don't let such a silly slugfest between rival telescope companies disturb or confuse or intimidate you. You could probably take the silly filberts out of the telescopes and not notice much of a difference. These companies are using their filberts for splitting very fine hairs, not double stars. And as far as the presence or absence of wertners are concerned, they didn't tell you that wertners cost only eleven cents each when bought by the bulk. Add to that that only eight people in the entire world have ever heard of Urbantke and Woolsey, and you will understand just why I'm urging you to simply let this roll off you like water off a duck's back.

Both outlets are large and successful — meaning that they deliver a very acceptable product commensurate with the price and the time taken to construct it. Don't expect state-of-the-art, because that's reserved for scopes that cost ten times what you're probably willing to pay. Barring the proverbial lemon, any imperfections in image quality will be negligible and visible only to someone possessing a very well-trained eye. If you happen to be one of the unlucky few who do get a lemon, grit your teeth, try to control your rage, and send it back. It may entail a lot of patience on your part, but the major companies will make things right, eventually.

But for now, make up your mind not to let their ads possess you. Remain the captain of your soul. Enjoy the ads, small, medium, large, and X-large. Appreciate the study and work that goes into them. Appreciate the money that goes into them. Order every catalog you can. The money you invest in such is money well-spent. It also helps the magazines because the telescope industry enjoys feedback and continues to encourage it by placing ads. The revenue from the ads helps the magazines which, in turn, helps you.

Just don't let the claims of the blazing ads blind you to what you really want: a good, useable, affordable telescope.

You've been warned about the ads.

Let's take some more steps into the jungle. Next stop: the telescopes. Come along. I've got you by the hand. You'll be okay.

5 The Telescopes

YOU'RE PROBABLY TIRED FROM STUDYING all those ads, aren't you? Let's make camp here, relax, and talk telescopes a bit. You need to purge your system of all the ad hype before you even consider buying a telescope. You also need to consider what type of telescope might be right for you.

Since you've hopefully done your homework concerning the ads and catalogs...

"But I haven't."

Let's pretend you have, okay? As long as we're pretending we're in a jungle, let's go the whole route.

I don't have to tell you that telescopes come in all flavors, sizes, prices. Some are pedigreed; some aren't; many would have you think they are.

From here, this part of the jungle looks a bit like the candy store of your dreams (or nightmares). Unless you happen to be Son/Daughter of Sheik, you will have to walk amid all the goodies and select one morsel.

So, what's it going to be? A big orange lollipop? A big licorice lollipop? A big grape lollipop? A Bill Blass designer bonbon? One of those expensive imported mints? A Mars bar? One of any number of jawbreakers? Or maybe a long peppermint stick? How about a package of the necessary ingredients and paraphernalia to make your own?

"Can I have a closer look?"

Certainly. Let's just browse for a few minutes.

"Look at all the refractors. Nice, huh?"

Yeah. Refractors look like telescopes, don't they? There are six different brands and a Heinz number of varieties in this fictitious "candy" store. Some look too skinny or short or both to be worth their price tags, don't they?

"All of them do."

I know. Good refractors aren't cheap, especially when their apertures hit four inches or larger.

"But this 4-inch alt-azimuth costs more than this 8-inch Schmidt-Cassegrain with a motor drive!"

Check out the price of the same 4-inch refractor with a motor-driven equatorial mount.

"Twice the price of the 8-inch Schmidt-Cassegrain. I can understand paying more for the refractor lens — since they are much harder to make — but why so much for the mount?"

Comes with the territory. The manufacturers know that anyone willing to spend more for a refractor will not quibble about spending extra money for the mount. This so, they do not seem hesitant to grab for all the gusto they can get. Put a 4-inch reflector tube assembly on the same mount, and there is no way they could sell the package for more than the 8-inch Schmidt-Cassegrain.

"What are these designer bonbons? So small, but so expensive!"

Maksutovs. There's a whole lot of flavor packed into those little things. That little 3.5-inch model there can outperform most quality Newtonian, Cassegrain, and Schmidt-Cassegrain reflectors twice its aperture when it comes to definition and contrast. The bad news is you can buy three 8-inch Schmidt-Cassegrains for what you will spend on that little beauty with less than one-quarter the light-gathering capability of the 8-inch. Some will argue that the increased definition and contrast is worth that sacrifice. I tend to agree — but then, I have been observing for a long while. I am immune to aperture fever and you are not. For now, pass these beautiful babies up. You can always come back later when you will be able to better appreciate and take advantage of what Maksutovs have to offer.

"Look at these monsters!"

Impressive, aren't they? A lot of poundage for the money, no? Those are 10- and 12- and 14- and 16-inch transportable reflectors, and they are priced less per inch of aperture than the Schmidt-Cassegrains. But do you have the muscle-power to enjoy them? Remember that even the Space Shuttle is transportable. You certainly don't want to get sick from eating too much candy in the game. Make sure your eyes are not bigger than your stomach.

"What are these great stubby, big chunks of stuff?"

Dobsonian reflectors.

''And this conglomeration?''

Telescope kits. Mirror grinding kits. Everything you need to make any type of telescope you want, provided you have the talent to mix everything properly into a digestible concoction.

''So much to choose from.''

And not an easy decision, is it? Let's spend a little more time considering what's available.

The Do-It-Yourself Kits

A little fast work with your handy pocket calculator will tell you that purchasing your own package of necessary ingredients and paraphernalia to make your own ''candy'' will not save you much money in the long run. Actually, it could end up costing you more.

''Why?''

Where telescope mirrors are concerned, new mirror-making techniques have brought the prices for finished mirrors down in recent years, while packaging all the stuff to grind them has sent the prices for kits soaring. Some firms offer mirror blanks alone for twice the price of another firm's identically sized, finished product. Granted, it doesn't make a whole lot of sense, but it's true, nonetheless.

Where mountings are concerned, if you will be content with a simple alt-azimuth — be it made of wood, pipe, or Tinker-Toy — you will save some money by making it yourself. Chances are you will not stay content for very long, as you will soon tire of constantly readjusting in both axes while tracking an object as the world turns. When that happens, the money you saved will haunt you, not please you. A good rule of thumb is to spring for the best mounting you can afford, whether you build it or buy it. Mountings are like tires. If you're traveling on good ones — no matter the vehicle — you feel better.

If you are mechanically inclined, have access to and an expertise in using precision lathes, drill presses, and such, then feel free to fabricate your own mounting.

Grinding your own optics is another matter entirely. Working with glass is not the same as working with metal. It requires a very special something. Glass is a merciless thing. It leaves you precious little room for error. It dearly loves to physically and psychologically torture those who do not have the knack of winning its affection. It can and will turn on you in the wink of an eye. It enjoys being fickle. It is unforgiving. And it breaks easily.

Glass will drive you bananas unless you have been gifted with the elusive magic touch of the optician. If you are so gifted, you will sense it. If not, likewise.

I wasn't gifted. I've known it from the very beginning. Not once have I ever even considered grinding my own optics. Glass does not like me. I've seen a 12½-inch mirror actually jump from my hands and self-destruct on my garage floor.

Why anyone would want to tackle grinding and polishing a mirror or lens is beyond me! Not when there are so many excellent ones available for such reasonable prices. Alas, you must decide rather early in the game if you do or don't wish to be an amateur telescope maker, an amateur astronomer, or both.

So search your soul some more. Do you want to observe or spend much of your observing time building something to observe with? Many ATMs spend years fabricating their instrument, then defabricating it to clean the "bugs" out, only to find that they've taken no time to observe. In fact, there are ATMs out there who could care less about observing.

Where I'm concerned, years spent toiling in some workshop are years I could better spend visiting the stars. I figure I've got a finite number of hours set aside for me to take in an infinite universe — and I don't want to give up one precious second of that allotted time. Thus, I am in no way, shape, form, or fashion, an ATM.

(Now, before a mass of livid ATMs swoon while reading such "blasphemy," let me stress that I know there is something very special about observing with a telescope and/or optics that one's own hands have formed. I do not begrudge that feeling of accomplishment. I stand in awe of anyone talented enough to pull it off. But for the tenderfoot, it's like downhill racing. From the spectator's standpoint, the extreme difficulty of each maneuver is elusive, almost hidden to the point that everything appears, on the surface, surprisingly easy. Suffice it to say that, if it were easy, telescope manufacturers would not be getting good money for their services.)

If you must start out in astronomy's jungle by making your own telescope and/or grinding your own mirror, start out small. Purchase an inexpensive 6-inch kit and give it a go. If you come up with a gem, fine. If you develop an ulcer, use the money you saved by not buying a bigger kit and see a doctor to get well.

There is one exception, however. Anyone with a thimble full of brains and talent can build a Dobsonian reflector.

The Dobsonian Reflector

At this writing, the Dobsonian revolution is still going strong. In a nutshell, a Dobsonian reflector is the simplest and best method for getting the largest aperture for the least amount of money. They are inexpensive light-buckets. Because of their ingenious design, these telescopes are very gentle on mirrors. This so, the mirrors can be thinner, thus less expensive than "normal" mirrors.

We have John Dobson of the San Francisco Sidewalk Astronomers to thank for this design. We're talking large aperture, stability, ultra-smoothness of operation (even though alt-azimuth). We're talking lightweight and ease of transportation for the aperture size. We're talking some gosh-wow views of the heavens.

Once you've got the mirror and diagonal in hand, the rest is almost child's play. All you need is some plywood, wood screws, wood glue, Formica, and a handsaw, a screwdriver, Teflon, some elbow grease, a few other incidentals — such as Richard Berry's *Build Your Own Telescope* — and you can make something like a 10-inch reflector for under $350. The only sticky part is obtaining the Teflon. Hopefully, by the time this book is out, most dealers who supply Dobsonian mirrors will also stock the Teflon. Right now, that's not the case.

If you are walking the Aesthetic Path, a Dobsonian may just be the only telescope you will ever need. After all, you want access to the beauty of the heavens, not a beautiful instrument.

If, however, you are walking the Scientific Path, the Dobsonian does have its disadvantages. Unless you want to hassle with installing a Poncet drive, the only means of driving the thing is by using your hands. It works this way, and very smoothly — but adjusting in both axes can prove aggravating while doing high-power studies of the Moon and the planets. The design does not cater to the demands of the amateur interested in seriously pursuing astrophotography, either. So, for the scientifically inclined amateur, I would suggest that the Dobsonian is a darn good choice for a second telescope, but not a first.

Maybe you get hives when you even go near a screwdriver. Maybe your mother won't let you play with a handsaw. Maybe even thinking about the sound grinding glass makes sets your teeth on edge. Maybe your toolbox is buried somewhere deep in the wilds of your storage closet protected by killer roaches. Maybe, just maybe, you don't have what it takes to build anything. Not to worry. I couldn't drive a nail if it had power steering — but that didn't keep me from becoming a survivor.

Okay, let's say you're going to purchase a ready-made telescope. In this imaginary "candy" store of astronomy's jungle, let's presume that there are a whopping 110 manufacturers represented.

For the sake of brevity, let's consider just a few fictitious examples: Precision-Eye Optics, Betelgeuse Optics, The Optimum Corporation, Starstep Incorporated, Copernicus Systems, Zanzibar Scientifics, and Yamato Enterprises. All totaled, these fictitious companies make thirty-three different models of telescopes which vary in price from $400 to $32,000.

Optimum has marketed their state-of-the-art 4.2-inch and 7.3-inch cadiotropics (mixed lens/mirror optical systems) for over twenty-five years, and their credentials are impeccable. Their least expensive fully-equipped model kisses $3000 — anything but cheap when you can buy a different cadiotropic design from Starstep or Betelgeuse or Precision-Eye for less than one-third that figure, and almost double the aperture to boot.

"A 4.2-inch for $3000 and an 8-inch for $825? Something is fishy here."

Now don't get paranoid already.

"But I've got the ads right here. Both Optimum and Starstep guarantee exquisite optics."

I told you to ignore the ad hype. Don't waste your time considering the claims. Instead, decide first the dollar amount you can most comfortably spend on your first telescope. Set $1000 as your limit, and you've ruled out state-of-the-art optics. Push that amount up to $3000 and you can get state-of-the-art, but the package will be small — $5000 will get you large aperture and an adequate mounting, but not state-of-the-art in either case. And I don't care how filthy rich you may be, spending over $5000 on your first telescope is dumb.

For now, just keep your checkbook safely tucked away while we continue looking around.

Since you're probably an average Joe like me, I strongly suggest setting your price ceiling around $1000. That cuts out Optimum. Don't worry about hurting their feelings. They know they'll probably grab you somewhere down the time line. It also cuts out the various companies who manufacture quality 4-inch or larger equatorially mounted refractors, and all non-Dobsonian reflectors larger than 10 inches of aperture worthy of serious consideration.

What's left for grabs? 17.5-inch or smaller Dobsonians (unless you make a larger one for yourself). Three-inch or smaller equatorial refractors. Four-inch or smaller quality alt-azimuth refractors. Ten-inch or smaller equatorial Newtonian reflectors. Eight-inch or smaller Schmidt-Cassegrains. There are, of course, other types of systems available — but the ones mentioned are those you will most likely consider.

Let's take a look-see at some of the possible choices offered by the fictitious manufacturers. First, we will consider those companies who specialize in fabricating one of the best-selling telescope designs of all time.

The Schmidt-Cassegrains

You've seen the ads. You can't miss the ads. Precision-Eye, Betelgeuse, and Starstep take out full- and double-page spreads each month.

Precision-Eye claims their Schmidt-Cassegrains are the most popular telescopes in the world, and who are you to argue with such a claim? But if you've even been to high school, I don't have to tell you what "popular" can mean. Let that useless bit of ad hype slide right off your back. They can't really prove that claim, and you can't disprove it.

Betelgeuse, not to be outdone, claims their Schmidt-Cassegrains are the most beautiful telescopes in the world. Granted, beauty sells, but just how important that is when you are observing in the dark is marvelous fodder for a debate team.

Starstep proudly admits they sell the world's lowest priced Schmidt-Cassegrains — but just how much lower those prices are remains to be seen, especially when certain hidden costs must be considered.

If the companies advertise their price for, say, their 8-inch models, that figure is almost always the base price. This will get you a tube assembly, fork mount, motor drive, manual slow-motion controls, setting circles, finder, carrying case, and one (possibly two) eyepieces. This will not get you a much-needed equatorial wedge or a tripod. Certainly, the scope can be used without these accessories, but only as an alt-azimuth (which makes the drive and setting circles of little use) that threatens to tip over if you breathe on it wrong.

To get full use out of such telescopes, you will have to spend more than just the base price. Unless you build the wedge and tripod (or pier), you will have to pay good money for these suckers. They aren't cheap.

Spread out your ads and catalogs and let's write down some prices on your trusty note pad:

	Betelgeuse 8-inch	**Starstep 8-inch**	**Precision-Eye 8-inch**
Base Price	$781.35	$779.00	$800.00
Eq. Wedge	50.00	55.00	50.00
Tripod	125.00	110.00	140.00
Totals	**$956.35**	**$944.00**	**$990.00**

At first glance, it would seem that Starstep gives the better deal, but the prices of these complete observing packages do not reflect their true values when all things are considered.

Precision-Eye's package is the highest, but they offer two high-performance, research-grade eyepieces in the deal. Starstep offers only one. Betelgeuse offers two, but theirs are not of the highest quality. Notice that their catalog says ". . . well-corrected eyepieces," not *highly* corrected. A little ad research will tell you that Precision-Eye's research-grade orthoscopics retail for around $35 each. Betelgeuse's retail for $14.50 each. And Starstep's one eyepiece sells separately for $39.50. That means the eyepiece packages are worth

$70, $29, and $39.50 respectively, which makes Precision-Eye's telescope package price a bit more palatable.

Note also that Precision-Eye's tripod, unlike the other two, is a variable-height tripod, justifying its higher price.

And there are other things which need to be considered. Betelgeuse charges $40 for packing. Starstep refuses to charge for packing, but their carrying case is made of sturdy fiberboard and not the luggage-quality material used by Betelgeuse and Precision-Eye.

Precision-Eye offers a twenty-five year warranty. Betelgeuse offers a five-year warranty on the drive only. And Starstep offers a one year unconditional money-back guarantee (if you can prove beyond a shadow of a doubt that you've never so much as bumped it during that time).

"There is so much to consider."

Hey, that's just the tip of the iceberg. These are big companies, remember. All three have authorized dealers around the world. Do you know what that means? It means that you just might be able to get a Betelgeuse 8-inch Schmidt-Cassegrain — complete observing package — for less than $956.35. Likewise on the other two companies. While the manufacturers hold firm on their list price, the authorized dealers do not.

Say that Betelgeuse discounts their telescopes by 20% to Click-Click Camera in Roanoke, Virginia. Well, you know of Click-Click; they take out that gaudy ¼-page ad every month. If their month has been a slow one, you just might haggle with them and talk them into settling for a quick 10% profit. Why should the guy turn you down when he can pocket the 10% and order another?

"How am I going to know about Click-Click (or a dealer like Click-Click) if they don't advertise?"

If they haven't already, each company will gladly send you a list of their authorized dealers. You will save money because the competition among authorized dealers is very keen. There is a catch, however. If you have troubles, you must deal with the authorized dealer. Don't scream at Betelgeuse if you get a lemon. They'll tell you real fast where to go . . . Click-Click. If the manager of Click-Click is reputable, you should have no problem. If not, he just might tell you where to go, if you know what I mean.

But perhaps you are one of those almost extinct creatures who will not be seduced by the full-page ads. Maybe you are leaning toward Copernicus Systems or Zanzibar Scientifics. They sell a line of Newtonian reflectors starting with little 4-inch rich-field models and going up to 12½-inch transportable reflectors, and they've both run modest ¼-page ads for years.

The Newtonian Reflectors

Like refractors, these babies look like telescopes, with their long tubes and black-crinkle equatorial mounts, neat guide telescopes, and little black power cords trailing down like tails. The photo-illustrations of them in the catalogs always make them look powerful. There's usually a very intelligent-looking gentleman dressed in sterile white lab coat standing by the scope and gently caressing the tube. You get the feeling such instruments are dust- and germ-free.

Let's take note of some prices:

Copernicus 8-inch f/6 (complete telescope): $595

Zanzibar 8-inch f/6 (complete telescope): $625.

Returning to the photo, the first thing you will think is: "Holy Erfles! You mean I can get that lovely thing that cheap?"

No. No way.

See that little bit of fine print in the upper left-hand corner of the photo? The one stating: "8-inch pictured equipped with optional setting circles, finder scope, guide refractor, rotating rings, motor drive, heavy-duty mount, and leveling screws."

"But it would look naked without all those things."

Precisely. That's why they include the stuff in the photo.

Again, let's do some figuring. Let's add up the price of those optionals:

	Copernicus	**Zanzibar**
Setting Circles	$25	$22
Finder Scope	$30	$25
Guide Refractor	$75	$95
Rotating Rings	$75	$65
Motor Drive	$85	$90
Heavy-duty Mount	add $65	add $50
Leveling Screws	$25	$20
Total	**$380 extra**	**$367 extra**

In other words, if you want the scope pictured in Copernicus' ad, it will cost you $975. Zanzibar's will cost you $992. And both charge another $25 to pack the scopes.

A similar story holds true for their 6-inch models.

There is really no sense in considering the stripped-down models. You've fallen in love, perhaps, with the one in the picture. All you would do if you ordered either one of the naked models is want it decked out. In a short time, you'd probably spend the extra money on the optionals anyway, especially if you are walking the Scientific Path. And too, if you should happen to grow disenchanted with the telescope and/or astronomy, a fully-clothed model is always easier to sell, and you will take less of a loss, if any.

"You never mentioned these inexpensive 4-inch models."

That's because I don't think you should start out with anything smaller than a 6-inch if you're going the reflector route. If you survive, in the future you will probably purchase a 4-inch rich-field, but that is a specialty designed to give you wide-field, low-power views of the heavens. For now, I want you to decide on a telescope for all seasons.

A good quality 6- or 8-inch motor-driven, well-equipped, sturdily-mounted, equatorial Newtonian is a very safe choice for your first scope. They are not as portable or as conducive to astrophotography as the Schmidt-Cassegrains, but they are not as expensive per inch of aperture and they are easier to align optically. Because their optical systems are easier to form, chances are you will receive as good or better image quality than from a Schmidt-Cassegrain of identical size, although the diffraction spikes from the Newtonian's spider may bother you. Newtonians come in lower focal ratios well-suited for deep-sky observation with a wider field of view than Schmidt-Cassegrains, and can easily be stepped-up with the help of a Barlow lens for planetary, lunar or double star work. Currents of air in the tube of a 6- or 8-inch Newtonian are not that bothersome, and neither is the diffraction caused by the spider. Overall, the relative ease of construction for the whole observing package gives the manufacturer less opportunity to botch the job and you a better chance of fixing something should the need arise. Compared with Schmidt-Cassegrains, you do give up portability, since sturdy equatorial mounts are no lightweights. Newtonians also cost more to ship (and *you* must pay the shipping).

I heartily agree with J. B. Sidgwick, author of the excellent *Amateur Astronomer's Handbook*. The advantages of a Newtonian reflector far outweigh the disadvantages.

"So I should buy a Newtonian, not a Schmidt-Cassegrain."

I didn't say that. We're not finished considering the possibilities. Consider. . .

The Refractors

Good old Yamato of Japan is the shining star here.

"There are no American manufacturers of quality refractors?"

A very weak yes. Remember now that we are talking about manufacturers who cater to the amateur *and* deliver a quality package priced within our $1000 ceiling.

Precision-Eye, in fact, is just beginning to market a line of what they promise are "affordable, quality refractors." But just how consistently "affordable" and consistently "quality" those refractors are going to be remains to be seen. A company called Faraway Scientific used to manufacture gems along this line, but a few years ago they were bought out by a larger firm, and, since then, the gems have lessened in quality. Xanadu Scientifics (remember these are all fictitious firms) markets a very pretty line of inexpensive refractors, but they are token telescopes manufactured for the sole purpose of lending credence to their ad claim, "We've got it all!"

Now don't get me wrong. There are American companies, both large and small, who do make quality refractors. They just do not cater to the amateur, especially where the prices are concerned. Those that *do* usually sell only the lenses and, perhaps, the parts to fabricate a tube assembly.

Fabricating a refractor tube assembly is not as simple as it may seem. You just don't stick the lens in one end of the tube and a focuser in the other end. The lens must be critically positioned and the tube needs to be fully baffled. The supply companies usually have all the aluminum tubing and focusers and lenses you need, but try to find ready-made, easily-installed light baffles. They're a rare breed.

"I notice here that Xanadu Scientifics' catalog says nothing about light baffles in their refractor tubes."

Hey! I'm proud of you.

"That means they don't have them, right?"

You can just about bet your grandma's dentures they don't. If they did, they'd brag about it.

See? You have to study the ads long enough to be aware of what they *don't* say. That's the only way you'll ever nail down what you're not going to get if you buy from them.

"Okay. Yamato's refractors do have light baffles. It says so right here on page 3. Four on their 3-inch, six on their 4-inch."

Notice that Yamato tells you how many. Check out Celestial Works' catalog. They sell refractors. What do they say?

"Our refractors are light-baffled."

That can mean just one baffle between the lens and focuser. As a matter of fact, it does mean just that. Rest assured, if their refractors included two baffles, they would reword that product description and use "baffles," plural, or "fully baffled."

"Why are baffles so important?"

They cut down stray light and increase contrast. And superior contrast is some of what you are paying for when you purchase a refractor.

"Speaking of paying for it, I notice that Yamato's 3-inch equatorial is $1250. That's above the ceiling you set for me."

I know. You should consider their four-inch alt-azimuth model. It's $995.

"But you told me alt-azimuth mounts are a hassle."

More so under Newtonians than under refractors. And Yamato's alt-azimuth is something else. Its slow-motion controls extend on flexible cables to just under the focuser, where they stay, no matter the position of the telescope tube. Tracking an object is so silky smooth and easy that, after a short while, it will become instinctive, second nature for you, like braking a car before a red light.

"But $995 for a 4-inch?"

The lens of the Yamato 4-inch will deliver images to rival any 8-inch Schmidt-Cassegrain or Newtonian selling for a similar price. For planetary, lunar, solar, and double-star study, it will satisfy even the most discriminating eye. It does remarkably well even in deep-sky work, thanks to its superb high contrast.

"That's hard to believe. An 8-inch has four times the light grasp!"

I know, I know. I know what the theories dictate. But the proof is in the observing. I have an excellent commercially-made 4-inch refractor, and it will hold its own against any commercially-made 8-inch reflector — barring state-of-the-art, of course. My refractor has taken on 10- and even 12-inch reflectors and come out on top where definition was concerned.

"You're trying to talk me into buying a refractor, aren't you?"

I guess maybe I am. I enjoy using my refractor. It never disappoints me. It's not as sensitive to atmospheric disturbance as larger reflectors are. I never have to take the time to allow its objective to "settle down" and reach ambient air temperature, unlike reflector mirrors. Like Schmidt-Cassegrains, it suffers from no image-degrading tube currents. And for all practical purposes, it is permanently collimated.

"But what about the color problem?"

Take my word for it. The much-talked-about and notorious chromatic aberration associated with refractors will not bother you, providing you've purchased a quality lens and take advantage of the quality-corrected eyepieces available today.

"Seems like I would be risking a very early case of aperture fever by starting out with a 4-inch refractor."

If you start out with a good quality 4-inch refractor, I am confident you will get hooked on refractors. The late and great observer John Mallas devoted his life work as an amateur astronomer to observing through his 4-inch refractor. Famed Walter Scott Houston, long-time writer of the "Deep-Sky Wonders" column for *Sky & Telescope* magazine, uses a 4-inch refractor for most of his observing. They, too, got hooked on refractors because refractors deliver. Aperture fever never bothered them that much because they were too busy enjoying their refractors to notice.

If you do get the fever, though, it will most assuredly be for another larger refractor. But Yamato's 5-inch model comes only with a super-deluxe, observatory-type, equatorial mount and costs a whopping $8500. $8500 is one of the quickest cures for aperture fever I can imagine. And anyway, once you've fallen in love with a quality refractor, you can always build yourself a cheap Dobsonian light-bucket.

"No matter how good the alt-azimuth mount on Yamato's

refractor is, you can't pursue serious astrophotography with it."

Good! You will have to be content with observing for a while. *That is the key to your survival.* In your first stages of being an astronomy enthusiast, you must observe, observe, observe. Don't think of anything else. Don't do anything else. Don't think of cameras. You start fiddling around in this jungle's photographic swamp this early in the game and the silver quicksand will get you real quick.

"You still haven't convinced me. I mean, a Japanese refractor?"

I didn't say you had to get a Japanese refractor. I simply said that Yamato has a reputation for making really nice ones. I have high hopes for Precision-Eye's new line. If there happens to be an authorized Precision-Eye dealer within driving distance of your home sweet home, and if you think you might want to go the refractor route, get the dealer to let you look through one. They may turn out to be the best thing since peanut butter and jelly sandwiches.

"If they're so good, why don't more amateurs pick refractors instead of reflectors?"

Beats me. I suppose because they've been seduced by large aperture optics. Perhaps the refractor's high price per inch of aperture has scared a lot away. Perhaps not enough of them ever got the chance to observe through a quality refractor. I dunno.

I do know that, barring a nuclear war or sudden ice age, a resurgence of interest in the refractor is most probable. People are slowly coming to realize that more light does not necessarily mean better image. They are growing tired of seeing "soft" stars in their eyepieces. They want crispness and clarity. What they want is the quality refractor image.

The new age of the refractor is about to dawn. I can see the first rays of its light just beginning to caress the horizon.

And so can Precision-Eye. That's why they're working late to be ready for the dawn.

"I still can't believe you're suggesting a refractor for my first telescope."

A quality 4-inch refractor on a quality alt-azimuth mount, complete with manual slow-motion controls. It will not disappoint you.

Later, after you have paid your observing dues, if you wish to get into astrophotography or feel the need for a motor drive, you can always add an equatorial mount. Then, you should have an observing package that should provide you with observing pleasure for at least 163 years. At the end of that time, check back with me, and I will give you another twenty-three years of work to accomplish with the same instrument. After that, you probably won't need me anymore.

No matter what telescope design you choose, remember to do your homework before you buy. Ignore the ad hype. If you have any questions about the telescope, write the company and ask them.

Be sure you are right, then go ahead.

Go deeper into the jungle. I've still got you by the hand. I can also tell you have telescope fever real bad. You better read the next chapter right now.

6 Read 'Um and Don't Weep

YOU ARE NOW WALKING DEEPER INTO the jungle where the beasties prowl. The fever has you. Something is percolating in the pit of your stomach. Sleep is something other people do. Your daydreams alternate from 40mm wide-angle to 4mm high-power. Every time you see a number, there's a shimmering "f" before it. And you know you're about to buy a telescope.

Take care. The jungle's beasties consider you fresh meat. But take heart — as long as you keep your head and check out the jungle's shadows carefully before you carry on, it will be worth every danger braved and every beastie fought.

I've had the fever myself for nearly twenty-five years. Now, six telescopes and thousands of observing sessions later, I can say that I have experienced first-hand most of the ups and downs and beasties this wonderful jungle has to offer — and lived to tell about it.

Believe me, the ups can take you as high as you want to go. The downs? That's another ball game; they can make a vegetable out of you. My right knee (the one I kneel on to sight through my telescope's finder) must be part cauliflower. And the beasties? They've pounced on me from every direction at times when I least expected them to. Several times their sharp teeth and claws have inflicted almost fatal wounds.

I truly suffered from exploring this jungle without a guide. But don't worry. I'm okay now. Though my sweet wife thinks I'm wacky to drive miles out in the wilds to seek a dark sky, an acre of ground without a single confounded vapor light, and a double star that probably won't resolve, I believe that I'm anything but wacky. So what if my focusing hand twitches uncontrollably whenever I hear the words "eyepiece," "drop," and "dirt" during the passing of a single day? And I try not to worry about my sanity when the words "eclipse" and "clouds" cause me to do silly things, like trying to direct-dial Poland or threatening to trade in my best telescope on a leaky bass boat and three weedless lures. I'm not wacky. I'm definitely not insane — I'm just crazy about astronomy!

Before you take one more step, read these survival tips.

1. Never Bite Off More Than Your Wallet Can Chew.

It isn't worth it. Buy a telescope you can easily afford. If the $1000 ceiling I suggested earlier is too high, lower it and walk stooped for a while. Bigger is better when it comes to gathering light, but not when it comes to observing with genuine peace of mind while you're up to your orthoscopics in debt. You can't really enjoy a telescope if you have to worry about the payments — or the vacuum it left in your bank account. If you're rich, this still applies: the telescopes involved are just more expensive, that's all.

2. Don't Let Telescope Fever Worry You.

Half the fun of owning telescopes is in the years spent wanting — and working up to — more expensive telescopes of different makes and designs. And half the surprise of owning telescopes lies in realizing that after you have worked up to what I call the Hernia State (10- to 14-inch *portable* reflectors), you'll find yourself browsing the magazine ads and catalogs in search of a smaller, more portable instrument just like the one you sold six months ago. Half the agony of owning telescopes is in finding that the smaller scope you now want costs 20% more and will take six months to complete. And half the mystery of it all is that, still, in the back of your mind, you would give your eyeteeth to own that 1220 pound 16-inch that disassembles in two minutes into sixty-three parts and fits into the glove compartment of a VW.

3. Do Some Hard, Realistic Thinking Before You Order.

You've already studied the ads, articles, photos, and photo captions in the magazines. You've already ordered the dealer's catalogs, devoured them cover to cover, and had the fine print for dessert. You've tried to ignore your racing heart, the dealers' sales pitches, most adjectives, and figures like 1200x. You've paid specific attention to the real telescope hardware descriptions, accessories included in the base price, portability, assembled weight (which is quite different from shipping weight), guarantees, and packing and crating charges.

You've written letters of inquiry to the dealers. As a courtesy, you have taken the time to mention where you saw the ad. You've talked turkey (I hope) about delivery times (which can run to more than a year), approximate cost of shipping, and the availability of optional accessories.

Now, I want you to get a little tougher. Take out the ads and catalogs again and inspect them with a fine-toothed comb. Check for tiny little critters that could in any way cause you trouble in the future. Chances are, if you look close enough, you will find some.

Case in point: Fictitious Zanzibar Scientifics markets economically priced 6- and 8-inch reflectors. Their ads photo-illustrate the 6-inch only, which is a very affordable $325. For the price, the 6-inch is not a bad buy. Certainly, the equatorial mount isn't large by any stretch of the imagination — but what can you expect for $325? The shafts of the mount look rather thin, just barely adequate. The ad does not specify their diameters (for good reason) but rock-like stability is assured (the size of the rock is not mentioned). The ad implies that large shafts are not needed for the ultra-light telescope tube (which they will tell you is made of spiral-wound cardboard, if you ask).

But hey, you've discovered all this, haven't you? You know good and well that they can't give you everything for such a cheap price.

The 6-inch is not the problem. The economical 8-inch is the problem. This bugger is infested with critters.

Zanzibar sells it for $550. It is not photo-illustrated in the ad, but only casually mentioned (almost as an afterthought) right under the bold price for the 6-inch. What they're saying to you subliminally is: If you're going to spend $325 anyway, why not plunk down just a little bit more and bag yourself an 8-inch! It's very tempting, especially when other 8-inchers on the market cluster around the $750 figure.

To discover the critters, read the short description of Zanzibar's 8-inch. In an off-the-wall way, it will tell you that the 6-inch and the 8-inch have identical mounts, save for a slightly longer declination axle, a heavier counterweight, and the addition of a manual slow-motion control and larger "rotating rings."

The reason they do not photo-illustrate the 8-inch should be obvious: the 8-inch tube assembly makes the shafts look like toothpicks! There is no way on Earth this mount, coupled with such a tube assembly, could give you rock-solid stability.

Another case in point involves possible changes made in the hardware since the last ad or catalog printing. Some time ago, I purchased a $900 4-inch refractor (not my present one). I thought I'd done my homework, but the critters got me anyway. The refractor was advertised to have internal light baffles — but when I received the telescope, they weren't to be found. The dealer informed me that his company had not included them for years even though they were still mentioned in the most recent catalog description of the refractor. Again, read the fine print. The dealers reserve the right to make price and product changes any time they see fit; those changes aren't necessarily improvements. So before you order, write and ask them if there have been any changes made in any of the hardware since the last ad or catalog printing. Get the nitty-gritty from them in writing, and keep it. That way, you are protected in case the company writes, "Oh, we don't include the mirrors anymore."

4. Prepare for That Precision-Eye.

While waiting for your new telescope to arrive, instead of nibbling your nails or unknitting your favorite sweater, get cracking. On clear nights, get outside with your red flashlight and the latest issue of your favorite astronomy magazine and get to know the sky. If it's raining, brush up on the history of astronomy — or learn how to spell (or pronounce) Betelgeuse. Find out how to test a mirror or lens; how to align, balance, and collimate; how to care for your optics, etc. Read everything you can about astronomy, observing, and telescopes. Then, read it again.

5. Know What You're Buying.

Study the classified ads in the magazines. ASTRONOMY and *Sky & Telescope* have them every month. There you'll find names and addresses of many people who share the same interests as you. Write them and ask for their opinions and suggestions. Include a self-addressed, stamped envelope as a courtesy. Often, lasting and fruitful friendships germinate in just such a manner. You may even want to purchase your first telescope second-hand. If so, you will find them advertised here: every shape, size, and model. Or you may want to inquire as to why someone is selling the telescope of your dreams? (If he or she hasn't sold the second-hand telescope in question, you may or may not get the real story. If the telescope has already been sold, you may get more than you bargained for, such as finding out from someone who knows everything the ads did not mention.)

Study the catalog picture and description over and over — not the ad, at this point. Don't be in awe of the instrument when it arrives. Know it as best you can before you get it. Then, when you unpack it and your living room is literally full of out-of-state newspapers and used masking tape, you will instinctively know that the finder goes there and the star diagonal clamps on here and something should be over there. If something is missing, you'll know and try to find it. (Experience has taught me that it is probably buried under at least two feet of wrinkled classified ads, wrapped in funny papers.) If, after searching each and every torn piece of paper, the missing object isn't found, you can inquire first with the shipping agent (especially in the case of concealed damage), then with the dealer. That way, you won't suddenly realize that just that morning, along with the unbelievable amount of packing paper, the garbage truck took your two finder brackets as well.

6. Get to Know Your Telescope.

Give yourself time to get acquainted with your new scope. And never, never be disappointed with your first view through it! It takes time to acquire the ability to appreciate the delicate wisps of starfog in something like the Lagoon Nebula. Just as you must learn to ride a bicycle or ski, you must learn to observe. Half the fun of observing is in realizing that you are getting better, that you can detect the greenish tint in that planetary nebula where before you had a hard time seeing a nebula at all. So don't let your first few sessions get you down simply because you can't see everything described in the magazine's observer's pages. You've got to work at it and earn your stripes — this is what separates the sheep from the goats in astronomy.

7. When You Show Your Telescope to Others, Be Gentle.

Don't expect someone unfamiliar with observing and telescopes to walk right up to your little 4mm eyepiece and spot those tiny craterlets on Plato's floor. Often, they'll cover one eye with one hand, squint as if they're about to get a flu shot, approach the eyepiece like it's going to bite, grab the tube with their free hand, kick the pedestal accidentally (for some reason, this always moves the telescope's line of sight to the only area in the entire universe void of stars), look at the glare of that ever-present vapor light in the lens, and just to be nice, say, "That's nice." From then on, they suspect that you have a very vivid imagination and that you should be placed in a home.

To prevent this, start them out with a wide-angle, low-power eyepiece. Buy the best you can easily afford, for you will use it often and come to love it. The eye relief is comfortable, the field bright and wide and obvious to the untrained eye. (Before you order, make sure it will fit the eyepiece tube of your scope.)

Be proud of your telescope and be eager to share it. You just might kindle an interest that will develop into a roaring devotion. The universe is something we all can share and still keep for ourselves.

8. Don't Try Everything at Once.

Earn your stripes observing before moving up to more specialized fields of work, like the aforementioned astrophotography. I wish that there was some unwritten law declaring that no person could even put a camera on his or her scope until he or she has logged at least 500 hours of genuine observing. There is so much to see — and so much to learn about how to see. It's very easy, for instance, after you get deeply involved in astrophotography, to slide into the habit of using eyepieces merely to locate and help track an object instead of to appreciate its beauty. This is sad. True, photographs are fantastic — and an integral part of astronomy — but they should never take precedence over those pictures that develop so softly, so well, so magnificently on the optic nerve of your eye. Of course, those pictures are never published, never impress friends or win awards, but they are always with you. They never gather dust or fade. And there's always that special pleasure in knowing that, when you look upon that distant light, it has traveled all those light-years — such an incredible journey — for you. Whenever I really contemplate that, I swear it brings tears to my eyes.

Now that you've read my eight most important survival tips, I will leave you alone for a while to make out your order.

Never fear. I shall return. There are more beasties lying in wait along the way. You will need some protection.

7 What Accessories Do You Really Need?

I HAVE RETURNED, AND I CAN SEE YOU.

You are standing before a mailbox, holding *the* envelope. For the twentieth time you check to make sure the envelope is stamped, correctly addressed, and sealed. Your heart is pounding like a kettle drum. Beads of perspiration glisten wetly upon your forehead.

Suddenly, you move the envelope toward the slot, but then you hesitate. For a second you wonder if you've made the right decision. Wouldn't geology be much less expensive?

But then, you think back a bit. You've done your homework.

Finally, after much soul-searching, you've selected a telescope that you, your wallet, and your back can live with. You've carefully filled out the order form and correctly computed the total cost of the telescope including packing and crating charges, if any. You've checked and re-checked the shipping address on the order form — each and every word and number is correct, including the zip code. You've even added your phone number, just in case.

So go ahead. It's time. Mail it.

There! It's done. You just ordered your first telescope. Congratulations!

Now for a few predictions:

From this day forth until delivery, worry will nibble away at your sanity morning, noon, and night. If you just stand there and wait, you'll go bananas in 6.23 days. Unless you occupy your mind during this agonizing waiting period, your nerves will be worth about the same as a 10-inch f/4 richest field mirror used as a moon hubcap on a '51 Chevy pickup.

Take my word for it, when the instrument finally arrives, chances are you will be well pleased and ready for clear, dark skies. During the next three weeks of the first monsoon rains your location has ever recorded, you will dust, adjust, align, re-align, and admire that precision eye, listen to the rhythm of the falling rain, hum the soundtrack to *Star Trek*, and calm down a bit. For hours and hours you will simply sit and stare at the culmination of your dreams. For a while, your life will be complete.

For a short while . . .

Like the call of the wild, accessories will start whispering to you, irresistibly seductive.

Yes, they will. And you will hear. You will listen. For like dust on a primary mirror, it is inevitable.

From the ads and catalogs, eyepieces, filters, astro-cameras, special films, guide telescopes, richest field finders, drive correctors, gargantuan star atlases, door-stopper textbooks, slow-motion controls, image intensifiers — even observatory domes — all will beckon.

You will try to ignore their hypnotic calls. But ever so soon, you will find yourself (only to pass the time, of course) making a short list of the "must have" extras. When you've finished, if you run true to average, the cost of these extras will add up to a shade more than the amount you will make during the next three years, before taxes.

Your list of desired eyepieces alone will be a veritable zoo of classy glass gems, medium-priced multicoats, loanable cheapies, and some "slightly worn" war-surplus Erfles. And after this, you might even find yourself checking real estate ads for a used but clean mountain top . . . .

What should you do, now that you've got Accessory Fever?

First: Come back down to Earth. It may be a cruel thing to make you do, but it's for your own good. There is no cure for the Fever. You will want all of those wonderful and even not-so-wonderful optional accessories, maybe even two of each. What now remains to be seen is whether you'll use common sense and slowly — but surely — purchase each item at the proper time.

But watch out: In your delirium, you'll come up with some great — even fantastic — excuses. I can hear 'em now:

1. I Need Every Star Atlas in Print, Because They're There!

Horsefeathers! (That's tough talk.)

If you aren't thoroughly familiar with the use of a simplified star chart the likes of which can be found in any issue of ASTRONOMY or *Sky and Telescope*, then it's foolish to be yearning for a more advanced atlas like the excellent *Norton's Star Atlas*. When you feel comfortable with the magazine star charts during actual observing sessions, then by all means go purchase *Norton's*.

Likewise, until you have mastered *Norton's*, you have no need of the Tirion *Sky Atlas 2000.0*. (Yes, I know it's impressive.) Consequently, there is little call to order Vehrenberg's comprehensive *Photographic Star Atlas* unless you are very proficient in the use of the *Sky Atlas 2000.0*.

You follow? All of these printed aids are fine and necessary tools of astronomy, but each one should be purchased in the proper sequence. The same goes for astronomy textbooks, guides, technical manuals, conference reports, advanced reference books, and the like.

2. I've Got to Have a Larger Finder. Mine Is Useless for Serious Work.

Serious work, huh? Come on, snap out of it! Try to remember that you've just received your first telescope, not a research grant. Don't let your head swell to such proportions that you'll need a counterweight just to walk straight. Counterweights aren't cheap. And remember, you've got the Fever.

For a while, at least, try to forget the words "serious" and "work" and just concentrate on enjoying your telescope. There is no law that says you can't have a little fun with it. Honest! And as for that small "useless" finder, have you really given it a chance? Bear in mind that some very experienced observers prefer zero-power finders.

Unless you know the night sky very well, a modest, low-power finder is the best to cut your teeth on. It will help your eye just enough, but won't overpower and confuse it with a plethora of new stars. At first, it can be very confusing to aim the telescope in the general direction of the celestial object you are seeking, only to peek through that large finder and see fifty stars where your naked eye saw five.

To make matters worse, that key star on the chart you planned to use as your crucial stepping-stone-and-get-my-bearings star is now lost amid forty-nine other gems. Right off, you'll begin to hyperventilate and scream. And, as every experienced observer knows, doing this near a telescope immediately fogs the objective, turns the seeing bad, and attracts night-raider ants to attack your midnight snack.

The whole thing is complicated even more if the finder is an elbow or right-angle type. Then your sense of direction can really go down the tubes because you are looking down, the telescope is looking up, and everything is reversed. Weird things happen. In your search you may get a red out-of-focus glow that you just *know* is either Mars or a red giant star with a pituitary problem. You eagerly move over to the main eyepiece and after you've focused out as far as your rack and pinion will stretch, find that you've locked onto a red aircraft warning light atop a TV antenna half a mile away.

This isn't to say such finders have no merits — of course they do. They're fine additions, especially to short-focus reflectors and catadioptrics. But you must know how to use them and that takes some time. Get to know your finder first, okay? Then try bigger and better things if and when you need to, not before.

3. I'll Need a Pack of Eyepieces. Different Guns for Different Game, You Know!

Take my advice, please. Do *not* sell your great-grandmother's priceless eagle-claw dining table (the one Ulysses S. Grant spilled turtle soup on) to finance the sixty-four eyepiece menagerie you've finally settled on. All you can do with sixty-four of the wretched things is wonder what to store them in, how to keep them in order, how to locate each one of them in the dark, and how to return home from observing with sixty-four instead of fifty-eight.

Face it, unless you're gonna use them for chess pieces, you don't need that many.

Granted, you will probably need a few more eyepieces eventually. When a dealer sells a telescope, he's careful to include in the base price those eyepieces (or eyepiece) that will give a modest and useable range of powers and keep the instrument's price well within the competitive range. This is fair and understandable business practice — it is, isn't it?

But until you've thoroughly used the eyepiece or eyepieces provided, you have no way of knowing what other ones you really do need. After fifteen or twenty careful observing sessions, you'll begin to see where your present eyepieces are good and where you could profit from another one or two. Then and only then should you consider adding to your collection.

When that time comes, think in terms of *quality*, not *quantity*. If you cannot afford state-of-the-art Plössls or orthoscopics, purchase quality Kellners instead. In the case of long-focus refractors or reflectors, Kellners often perform as well as orthos and are considerably less expensive. In any case, make sure they are the right size to fit your focuser.

And take it from experience, you can wait a while for that high-powered 4mm eyepiece. Rarely will our atmosphere permit you the pleasure of doing more than taking it out of its warm little nest and putting it back in (probably in the wrong slot). And anyway, you'll soon grow out of thinking of your telescope as a high-magnification instrument only.

Along with the new eyepieces, purchase the best Barlow (negative) lens you can find. These lenses amplify the power of the eyepiece they are used with. Before selecting such a lens, pay special attention to the clear aperture of the achromatic lens and its amplification. The larger the aperture, the better (as long as it will fit the drawtube of your focuser).

These lenses are readily available in amplification powers of 2x, 2.4x, and variable power 2x-3x. It is very important to select a Barlow lens that will not waste your eyepieces' time. For instance, if your eyepiece collection includes an 18mm, a 12mm, a 9mm, and a 6mm, all you'll get with a 2x Barlow is a seldom used 3mm eyepiece out of that 6mm. With the others, you are simply duplicating the powers you already have (albeit with some added eye relief). A 2.4x or a variable power Barlow lens would be the better choice in such a case.

A good selection of eyepieces for a 2x Barlow lens would include a 20mm Erfle, an 18mm, a 12mm, and a 7mm. The 2x lens then gives you the equivalent of a 10mm, 9mm, 6mm, and 3.5mm set. No duplications.

I know you've probably heard horror stories about Barlow lenses. It seems that when combining negative lenses and eyepieces, the resulting marriages are not always made in heaven. When a lens of poor quality is coupled with a quality eyepiece or vice versa, only one thing can happen — incompatibility.

Even the best negative lens will make an eyepiece of poor quality seem twice as poor. And the night's rotten seeing will probably do the rest. Divorce will soon follow and chances are it'll be the Barlow lens, not the lousy eyepiece, that gets put out to pasture. What a waste.

Remember, in optics, quality should marry quality.

4. I've Got to Have an Astrocamera. Next Month, I Want to See My Astrophotos Glutting the Magazines!

There you go again . . .

You've already got a priceless double-lens camera in your head — your eyeballs. And what about that fantastic darkroom you couldn't live without — your brain? Have you given them a chance?

I have logged well over two hundred eyepiece hours observing the Dumbbell Nebula alone (I think I relate to its name). Yet I will never photograph it. I know it too well. The photo would only be anticlimactic. It just wouldn't do the nebula justice.

The picture in my mind is a 200+ hour exposure, unblemished by reciprocity failure or film grain. Of course, I can't show it to you, but you can have your own version, more or less like it, with only a bit of effort and dedication. Just get outside with your telescope and start the exposure. Observe and study.

Observe and study.

There will be plenty of time later on to buy a camera and begin the trials, pitfalls, and tribulations of astrophotography. At this tender stage of your astronomer's life, why aggravate yourself with underexposures, overexposures, drive errors, and the like? Try observing — seeing the heavens' wonders before you even consider photographing them. And by observe, I mean OBSERVE. Quick thirty-second glances don't count. Only hours at the eyepiece can train your eye to see.

When you know beyond a doubt that you are ready for an astrocamera, realize first they they are not inexpensive, and that you will have to purchase certain extras if your camera is to pay off with beautiful astrophotos.

Unless you are content with short, unguided exposures of star trails and the Moon, you will need a quality equatorial mounting and motor drive. You will need to know how to align your telescope accurately on the pole, balance it properly, correct inherent drive errors, and most important, how to find pleasure in such activity.

If you plan to photograph deep-sky objects, you will need a variable speed drive corrector plus some form of slow-motion control on the declination axis of your telescope. Simply having a drive corrector doesn't mean you will be able to guide your telescope perfectly during a long exposure. The corrector adjusts for errors in right ascension, but not in declination. Some companies offer everything you would need in the way of variable speed drive correctors, but do not offer any slow-motion controls for the declination axis.

Even if your new telescope has a declination control, it may not be slow enough or smooth enough for long-exposure astrophotography. If such is the case, you will have to replace it. Some commercial telescopes come equipped with an electric declination slow-motion designed primarily to slew the telescope and/or help center an object in the field of view for visual observations. Some have a definite lag in their control response. Such controls are fine for cruising the stars, but are of little help in astrophotography when you need to nudge that teensy guide star just a tad to bring it back under the cross hairs of your illuminated reticle guiding eyepiece.

Take into consideration also that when and if you decide to really get into astrophotography, you'll be committed not only to fabricating or buying an astrocamera, camera adapters, guiding eyepieces, guide telescopes, and extra counterweights, but also to working with special films, learning special developing procedures, and mastering special photographic enlargement and printing techniques. Thus, a well-equipped darkroom begins to loom large in your plans.

This, of course, takes time and money. Very soon (in fact, appallingly soon) you can have more money sunk into photo-accessories than into your telescope. That once-modest, basic instrument abruptly represents a sizeable sum. What's more, you've been so busy creating the perfect setup that you've had no time to observe. You've missed the fun, the glory of it all.

Suddenly, you find a great responsibility resting upon your shoulders. And it weighs heavily. If you are to make that instrument pay, you've got to come up with some stunning astrophotographs. And that, you'll quickly learn, ain't so easy, friend.

Be advised: Those beautiful photos that so often adorn the magazine pages did not come easy. Beneath all of that infinite beauty and grandeur lies hidden a good deal of the photographer's "blood, sweat, and tears." For this reason alone, you should not merely glance at the photos. Appreciate them. Take the time to read the photo credit. Each of those photos is the work of some devoted, caring astronomy enthusiast who's paid his or her dues — and invested lots of time, effort, and money in perfecting his or her work.

Everyone must pay their dues, even you. But luckily you don't have to pay them all in one lump sum. Before you spend more on a complete astrophotography setup, be very sure you have taken full advantage of your present outfit. Astrophotography and its accessories are an exciting and integral part of the enjoyment of astronomy, but they are definitely not for the novice.

Be sure you're ready, then go ahead. But tread softly. There are a lot of hungry beasties in this section of the jungle.

5. I Can't Help It. I Want Everything. Everything!

And, as you've no doubt already noticed, everything costs money. But when the Fever hits you, monetary objections melt away. "Money is no object. I need it," you think. You haven't even got an A.A. (Accessories Anonymous) to get you through those long, dark, lonely, equipment-starved nights.

If you don't keep control of yourself, you will order away like a lunatic. In one fell wild swoop, you will deluge the mails with orders for safe Sun screens (and you're not even that interested in the Sun), nebular filters, Moon filters, accessory cases and trays, tele-extenders, telecompressors, electronic timing devices, every book in the book, rechargeable energy packs, power converters, dewcaps, personalized astronomical flashlights, automatic trackers, hydrogen-alpha filters (the Sun again), rich-field conversion units, and even digital right ascension and declination readout consoles.

Why, you can even purchase a complete television rig that will allow you to observe in the comfort of your own den, proud as a peacock, in living color. Every month, new, exciting, and innovative items come on the market. Does this mean that astronomy has sunk to the level of a "status sport" for the affluent? Luxury observing has definitely arrived — where will it all end?

I can already see the next step: Rent-an-Observer, Inc. This company will, for a nominal fee of about $129.95 a night (half-price if it's cloudy), rent you a fully-equipped human being who will take your telescope out to your favorite observing sight, observe and photograph the four (4) deep-sky objects of your choice, and return your telescope to your storage area at sunrise.

Just think: You won't have to fight the cold, dew, bugs, the misalignment problem, or your runny nose. You can even go out to the movies! Won't that be great?

Let's hope that never happens.

Let's never lose touch with the real pleasure of amateur astronomy — the observing, the appreciation of our limitless universe and its unbelievable beauty. That's what's really important; all the gadgets and paraphernalia should only be a means to this end.

Most telescope accessories are truly useful items and have their place in astronomy. But never let carelessly bought accessories turn a practical telescope into an overgadgeted, overweight monster.

If you take care, use your good sense, be patient, and just let things happen gradually, you won't create that monster. You will not become an astronomical Dr. Frankenstein. If you are walking the Aesthetic Path, that chance is slim. Not so if you're walking the Scientific Path.

Take it slow and easy.

Be gentle with your new telescope.

Be gentle with your funds.

Be gentle with yourself.

And observe, fellow astronomer, observe and enjoy!

8 What Type of Astronomer Do You Really Want to Be?

NOW THAT YOU'VE COMMITTED YOURSELF, you need to consider what type of astronomer you are going to be. Even if you are walking the Aesthetic Path, you need to be aware of the types, just in case you decide to change boats in midstream.

Of course, professional astronomers get the most press, with successful amateur comet hunters coming in a close second. Very rarely do they dominate the front page news, however. Sometimes I feel that were a professional astronomer to discover that Mars was made of peanut brittle, the story would be buried somewhere in the newspaper's food section.

Carl Sagan is probably the most famous and infamous professional astronomer of this century — famous because his book *Cosmos* became an international bestseller, with the Public Television video version picking up a respectable, though not record-breaking, audience; infamous because many stuffed shirts in the field believe that he is *first* a professional entertainer who has used the hallowed science of astronomy for, brace yourself now, financial gain!

The unarguable fact that Dr. Sagan has been very successful in opening up the universe's mysterious vault to the general public has been both a blessing and a bane for him. He has earned the love and respect of many, but also the scorn of some of his scientific kinsmen, who for the most part have been set upon by that green-eyed monster called Envy and goaded by that serpent called Jealousy.

"Why?"

For a long time, professional astronomy has been and still is a notoriously low-paying profession. There is no sense in embracing that life if you've got dollar signs in your eyes. This is a field where many astronomy Ph.D.s are driving cabs or the like because the job openings are just not there. Every professional astronomer holding down a job seems to live to about 93. It's the love of their work that keeps them going — that and the fact that their cholesterol level is usually very low since they can't afford or do not take the time to eat a whole lot. And too, living up on high mountain tops relatively free of pollution probably has a lot to do with their long life spans.

Most professional astronomers are not poverty stricken, but very few are millionaires. This so, I think you can understand why someone who sifts gold from the stardust just might be a prime target for scorn. Sagan struck gold, and in doing so, stepped on quite a few intellectual toes without meaning to.

"Well, bully for him! I couldn't be happier for him. I just hope he's been able to let the jealousy and envy run off his back and into the gutter where it belongs."

He's earned his money, for sure. He's gotten "billions and billions" of people to look up and think, "Wow, that is *something* up there!" He got astronomy noticed like it has never been noticed before. He explained it in a language everyone could understand, and did astronomy an invaluable service.

Things are different now. Here is what I mean:

Man/Woman-on-the-Street Interview — B.S. (Before Sagan)

Interviewer: "Would you both describe a professional astronomer?"

Man-on-the-Street: "Sure. They always wear baggy pants. All of them wear glasses, usually the rimless kind, sometimes even monocles. Those over forty haven't bought a new suit since WWII. None of them are into jogging. In fact, they never exercise. They know the names of every star in the sky. They all are good at math. I don't think many of them are married because they're too busy at night for . . . you know. They never tuck their shirts in properly. And they always wear these ties that don't match. They never, ever laugh. And all of them smoke pipes. When they talk, they frequently use words that would tie my tongue into a granny knot if I tried to say them. All of them are geniuses — but I wouldn't want my daughter to marry one."

Woman-on-the-Street: "I've never seen one."

Man/Woman-on-the-Street Interview — A.S. (After Sagan)

Interviewer: "Would you both describe a professional astronomer?"

Man: "Sure. Most wear turtleneck sweaters. They are relaxed looking — a lot like writers. They look smart. You can tell they never had to cheat on physics tests in school. They can tell you what a planet is made of without ever having been there. They work for peace and not for war. And their favorite number is billions."

Woman: "Sexy in an intelligent sort of way. You know, like Carl Sagan."

So, if you think you're "sexy in an intelligent sort of way," you might want to consider professional astronomy. Just don't do it for the money. Let's face it, there aren't many of us around with the charisma of a Sagan, or with the ability to coin a phrase and cause it to be printed on thirty-two million T-shirts.

If you think you might be willing to work this "hobby" professionally, check out the job opportunities in the academic, federal, and business and industry areas. Realize that you will probably have to earn your doctorate and still have trouble finding a job, then weigh the alternatives. You may want to pursue some kind of related astronomical occupation such as planetarium work, optics, assembly line work for a telescope manufacturer, sales, or my profession, astronomical science writing. And let me warn you right now, should you be toying with the writing end: I've eaten more than my share of cold pork and beans. Writing about astronomy is very rewarding, but hacking out romance novels is much more profitable, unless you have the talent of Sagan. I'm afraid I do not.

Now that we've dealt with the paying end of astronomy, let's consider the amateur side of the coin. You can be one or more of the following types:

The Gadget-Mad Amateur

This feverish individual never has any time to observe because he's always too busy tightening screws and wing nuts, adding expensive and exotic accessories to his already overloaded telescope, and negotiating overseas loans to finance his next necessary addition. His most prized possession is his state-of-the-art set of Allen wrenches. He has a screwdriver for every known screw head in the universe. He is not satisfied until his telescope is so covered with optional extras that the tube assembly is invisible. Every optional accessory has a back-up, just in case. NASA calls this "redundancy." There are some gadgets on his scope that he will never use: a geiger counter to warn him of radiation in case of a nuclear attack while he is deeply involved trying to find the scope's eyepiece; a flare gun just in case he gets lost in his work; an inflatable raft to protect him from floods or tidal waves; and a battery-operated television just in case he gets bored.

He's not only equipped the poor telescope with a compass, but when the polar axis is properly aligned, a tape recording goes into action that shrieks, "By George, he's got it!" The lenses of his four guiding refractors have tiny windshield wipers. The scope even has its own custom-designed license plate. The scope's finders have finders. And down near the bottom of the tube, near the mirror mount and Pepsi dispenser, sits a little canary in a gilded cage — this in case of poison gas. When not in use, the telescope has its own aircraft warning lights.

When this gadget-mad amateur first ordered his telescope, it weighed 95 pounds. Now it weighs 753 pounds. Its polar shaft is beginning to wilt. The drive gear had 365 teeth; now it's got 73. The worm gear wishes it could die. The poor mirror doesn't know what is going on, since something above the spider is blocking out all incoming light. That something is a tape deck bought from K-Mart.

When asked what kind of images his telescope delivers, our gadget-mad amateur will say something like: "I dunno. I don't plan to observe with it until I get it fully equipped."

This guy is in no way an amateur astronomer. He's the type of guy who would put earrings on Bambi. He would be just as happy if the universe did not exist — that way he wouldn't feel obliged to keep that darned tube assembly that takes up so much useable space.

When you call up an amateur like this and say, "I've just detected a supernova in M-51!", he's likely to answer, "I can't look at it right now. I'm baking a German Chocolate cake in my scope's microwave oven and I don't want it to fall."

I'll let you in on a suspicion of mine: gadget-mad amateurs must be members of some strange religious cult that worships the god, Doodad. Doodad's main claim to fame — if you will remember your junior high lessons in Warped Mythology — was that he shouted, "If I'd have made the universe, I wouldn't have made it so darn plain!", right before they took him to the funny farm for lunatic gods.

The universe didn't think too highly of crazy old Doodad; she chewed him up and spat him out.

If you want to follow in his ridiculous footsteps, go ahead. You will go, however, without little old me. I may be crazy — but not that crazy.

The Female Amateur Astronomer

If you happen to be female, for heaven's sake don't let that stop you from giving astronomy a try. Just accept the fact that you will probably be in the minority at any given star party or astronomy club meeting. You may even be the Lone Rangeress. You will get your share of stares, and you will probably have to deal with one or two clowns who still think that women should stay "barefoot and pregnant."

Yes, Virginia, rest assured that there are lassies interested in amateur astronomy. At first glance, it does appear to be a male sport — but that is not the case.

Females are much more suited, actually, to long periods of observing. For one thing, their lower center of gravity makes it easier for them to peer into an eyepiece for extended periods of time. Females are not as likely to kick their telescopes to smithereens (wherever that is) because of a sudden overcast of clouds. For some reason, female hands tend to stay cleaner — which can't do anything but help their eyepieces stay likewise. For cold nights, don't forget that females have that insulative layer of fatty tissue that men don't have. And too, I have it on good authority that the universe is a *she*.

So walk with your lovely head held high. Show everybody that you are ready, willing, and able to survive on your own, and the male amateurs will treat you with respect. Who knows, you may find that a surprising number of gentlemen will be more than willing to help you set up, hand you eyepieces, loan you eyepieces, bring you coffee, loan you film, develop your film . . . .

If you're not lucky enough to be near an active astronomy club or at least a group of enthusiasts who observe regularly, and if you happen to have a husband or boy friend who could not care less about starstuff, I strongly suggest you take a few precautions:

- Before you consider a telescope, know the limits of your lifting capacity. Some 8-inch reflectors are tests for lumberjack backs.
- Never go to a remote observing site without a sizable companion unless you're an expert in the martial arts. Observing alone out in the wilds can be a test of courage even for Daniel Boone-types.
- When you observe out in the wilds, don't wear perfume or

hair spray. They will mask the clean, invigorating fragrance of country air and draw bugs, too. And don't glop your eyelashes with mascara that will, in the blink of an eye, gunk up your eyepiece lens.

• Don't wear dangling jewelry. It will drive you crazy jingling against the tube as you focus, not to mention the damage you can do to a fine eyepiece lens.

• Do wear comfortable, loose-fitting, rugged attire, and forget the designer stuff. Besides being constricting, designer slacks will quickly sprout holes in the knees, for you spend a lot of time on your knees out there doing things like peering through finders, crawling around in the dark in search of the eyepiece you dropped, and praying that that one, tiny cloud covering Saturn will go somewhere else to play.

But, perhaps, most important of all is to not be self-conscious about being a female.

So come on, ladies! Astronomy needs you.

The Shutterbug Amateur

Here we have a species of amateur astronomer that only worships the Orion Nebula if it is captured on film. Pinpoint star images in a 20mm Erfle eyepiece do not get this individual's juices flowing. Pinpoint star images on a photographic emulsion, on the other hand, give him heart palpitations and cause him to drool.

The true shutterbug amateur astronomer couldn't care less about using his eyes for anything but focusing on a ground glass camera screen and keeping an out-of-focus guide star centered under the cross hairs of his guide telescope's illuminated reticle eyepiece. He's the guy who is always first in line at the Quik-Photo booth. He's got albums full of 8x10 photos of white dots on black backgrounds. These thick albums are usually placed someplace very accessible — most often, stacked high on his coffee table in his living room. His den is plastered with framed photos of what look like fuzzy tennis balls. He tells everyone that these are his best photos of Jupiter. His neighbors wonder why this funny man takes pictures of fuzzy tennis balls and then names them so strangely.

His goal in life is to sneak a picture of Saturn when it is least expecting it. He's taken 2000 pictures of that planet and never seriously studied it through the eyepiece once. If the pole star were to fade to invisibility, shutterbug would be lost.

Though he can tell you the characteristics of every film made, he cannot tell you anything about Jupiter's atmosphere or telescopic appearance, the reason for Mars' ruddy complexion, why a comet's tail always points away from the Sun, or what in the deuce M-42's Trapezium looks like.

Shutterbug bought his telescope because it was expressly designed for astrophotography, which means the company makes around $32,000 worth of optional, high-profit photographic accessories that fit it like a glove. And you can bet he's purchased everything including the kitchen sink.

He's good at what he does. He takes a really sharp photo every now and then. The various magazines covet these photos because they make for good, exciting visual graphics.

But Shutterbug is not an astronomer.

He's paid his dues, to be sure. He can tell you about frost-bitten fingers, dry-ice burns, brittle frozen film, double exposures, and clicking away night after night to average just five good exposures out of ninety. He can tell you about feeling the tears from his eyes track down his cheeks while he was guiding during a fifty-minute exposure.

I must say I respect Shutterbug's interest in astrophotography. It's just that I pity him. He is missing so much by not being an observer first. Astronomy is a love affair with the universe, not with photographic emulsions and cameras. The best astrophotographers today were expert observers first. They work to further astronomy by seeking to perfect new and easier methods of capturing significant changes on film — something that will increase our knowledge of what's happening up there, not something pretty that will look good on a den's wall. When a true astronomer/astrophotographer takes a photo of a distant galaxy, he/she studies the image to detect, perhaps, a supernova, or something not captured before, or a subtle change. It is not a game with them, but research. Certainly, they have fun doing it. And certainly, the results of serious astrophotography are impressive. But the big goal should not be the picture framed. The goal should be to discover what secrets the photo might reveal to one who knows well the object photographed.

Mr. Shutterbug has no desire to unveil any secrets. He's an

aesthetic astronomer walking the Scientific Path. He's where he should not be.

Never take the precious time to photograph an object until you've studied to the best of your ability what is known about the object, and studied its visual appearance thoroughly. A good portrait photographer must know his subject. Only then can he bring out the subject's true character.

If you want to be a shutterbug, then *be* a shutterbug — but be an observing shutterbug.

The Dyed-in-the-Wool Do-It-Yourselfer

There are amateur telescope makers, and there are amateur telescope makers. Believe me, there are ATMs out there blessed with the ability to design and construct anything their hearts desire. They will work sometimes for years on their home-built telescopes and create absolute gems of perfection that would make any astronomer salivate in his eggnog.

I wish I could do the same thing — but I have trouble putting together those 29-cent dime-store balsa wood gliders.

Some of these guys are only machinists or inventors at heart — but for the most part, they are also accomplished astronomers to the point of being professional minus the degree and job.

They can, however, go overboard.

Consider the Ultimate Do-It-Yourselfer: This person thinks there is nothing impressive about a telescope unless its owner hand-formed it out of crude ore and sand. He's the type who whittles his own ball bearings and buys used lathes from shut-down car plants. His is the telescope with the steam-powered drive. His spare time is spent grinding excellent eyepieces from broken Coke bottle glass. He sees no reason why a useable worm gear cannot be fashioned from a petrified carrot. When he shows you his Dobsonian, if you remark at the beauty of the hardwood tube, he says proudly, "I felled the tree."

Ask him where he got his mount for his home-built 8-inch refractor, and he'll say, "I got the aluminum ore from a mine in Virginia, the brass from . . ." You get the idea, don't you? Whatever you do, don't ask him how much that beautiful 8-inch refractor of his cost. He'll say something like, "Seventy-three dollars . . . in gas."

That'll make you sick.

If you can hold your own among such company, it is an enviable position. Just don't lose touch with the prime directive: Observe, observe, observe.

The Aesthetic Amateur

This type is special, no doubt about it. They must be given special treatment for they remain rather tender throughout their astronomer's lives. They walk to the beat of a gentle drummer. They listen to the music of the spheres in a relaxed, laid-back way. For them, astronomy is as balm. They do not seek answers; they seek peace and tranquility and beauty. They seek celestial poetry. They taste the wine of darkness and light, sometimes sipping it in moderation and sometimes taking in great draughts. They demand less of themselves and their telescopes. They tap into a special portion of the universe that is forbidden to those walking the Scientific Path. Religious experiences are not unheard of here.

For the Aesthetic Amateur, magic is real and attainable.

The Casual Scientific Type

This type is made up mostly of good Joes. They take things as they come, are usually open to many areas of interest, and have great fun trekking with the stars with their modest equipment. They do not bury their heads in technical journals, or focus on only one aspect of astronomy — but seek to take all of it in and enjoy the whole ball game, aesthetic astronomy included. They dabble in photography, double-star work, deep-sky observing, planetary study, meteors, comets, and good, non-technical books on just about anything concerning the universe. They usually have a Sagan T-shirt, and they wear it and wear it and wear it. They are not embarrassed to say, "I don't know how deep Jupiter's atmosphere is, but I'll learn how deep it is later." They keep on keeping on until, lo, after a few years of casual enthusiasm, they wind up knowing a heck of a lot more than they think they know. And what's more, they do not tend to continually slap you in the face with their knowledge (like I'm doing right now).

The Amateur with No Funds

Yes, there are those who cannot afford to purchase even the least expensive telescope. And yet, the universe has touched them, called them, plucked at their heartstrings, opened a very special door. These people walk the Earth at night, eyes sparkling under clear skies. They observe in the only manner they can — with their naked eyes. And they see more than many who are more fortunate. They observe.

***

So what type of astronomer will you be?

Only you can answer this question. You are the puppet and your own puppeteer. I'm willing to guide you, but I will not pull your strings.

9 Seeing What You Observe, Observing What You See

THE LIGHT COMING FROM A DISTANT celestial object must tunnel through extragalactic and/or galactic debris, trash, gas, and dust; drill some kind of hole through our filthy atmosphere; bend through the eye's lens; swim through the eye's jelly-like humor; and finally end its journey on the rods and cones of the retina if it's going to benefit the observer at all. Making it through all that stuff and still surviving the trip is no child's play — but surviving the translation after that is no vacation either, especially if the brain translating that feeble energy is ill-tuned, insensitive, and untrained.

There are two important types of seeing: one controlled by the conditions present in the Earth's atmosphere; the other at the mercy of the observer's ability to see what light is observable. We have little control over the former, but we can do something about the latter. With patience and persistence and practice, we can train our eyes more every time we observe.

For the average enthusiast, the first look through a telescope delivers a great deal less than expected.

"Why?"

Because most have preconceived notions of what they will see which are the results of seeing photographs of the celestial objects in question. A novice will expect old Jupiter to appear more detailed and larger than those photos taken by even large telescopes. When they peer into their eyepiece and see nothing but a diminutive, glaring ball of light, they will shrug their shoulders and think, "Big deal!" Chances are they won't even notice Jupiter's moons, because the moons do not look like moons at all, but stars. If this is their first reaction upon viewing through a telescope they paid good money for, this can really hurt. Immediately they will think, "What have I wasted my hard-earned money on?"

This initial shock of discovering what is really your initial inability to see is the reason I urge you to beg, borrow, or steal some viewing time with another person's telescope before you purchase one for yourself. If that is not possible, pay extra close attention to what I'm about to say.

You must give yourself time to learn to see. Viewing through a scope for the first time is like walking from a brightly lit room into a dimly lit room. At first, your eyes cannot see a darned thing — or, at least, precious little. They have not adjusted to the new light. If you turn around disappointed and walk back into the brightly lit room, your eyes will stay sensitive to only the light in the brightly lit room. If, instead, you stay in the dimly lit room for a while, your eyes will gradually adjust and you will see more and more as time goes by.

The same is true with observing — only the time involved in learning to see involves not minutes, but hours that grow into years.

First, train your new telescope on the Moon. Force yourself to observe only the Moon for a while. Exercise the limits of your vision until elusive details begin to become not so elusive: rilles, tiny craterlets, subtle undulations of the lunar maria, textures of crater floors, etc. Gradually, you will find that, with each observing session, more and more aspects of the Moon's complexion will present themselves to you. Also, you will find that on some nights these details remain invisible due to atmospheric turbulence or a change in its transparency.

The more you observe, the more you will understand the effects both types of seeing have on the celestial object observed. Atmospheric turbulence will rain on almost any observer's parade, whereas low transparency will bother the deep-sky specialist more than the planetary and lunar observer. In some cases, hazy conditions resulting in low transparency can actually act as a filter to improve planetary and lunar images while almost absorbing the useful light of deep-sky objects. Haze is a symptom of stagnant or stable air masses; these often provide very steady seeing. If the celestial object can push enough light through all that garbage, its image is often crisper than one would imagine. On the other side of the coin, the study of nebulae and galaxies mandates as much light as possible (high transparency) passing through the atmosphere and the eyepiece. Steady seeing is not so important in this case.

After you've got your feet wet observing the Moon, try out the planets. Jupiter, if it's up.

"It is."

It is because I'm in control here. In real life, it might not be. If so, try for Saturn. If *it's* not up, don't try Venus. You're not ready for the disappointment Venus will probably offer. And the same goes for Mars. If it's up, leave it for later. I know you won't — but at least you can't say I didn't tell you so. Should Jupiter or Saturn not be available, check out some of the brighter stars and star clusters easily spotted in your finder. Cut your teeth on them for a while.

But in this book, Jupiter is available. Zero in on it.

"Okay. Got it."

Now focus the image carefully until its moons are little dots or pinpoints, depending on the size of your scope.

"Done."

What do you think?

"Great!"

Come on! What do you *really* think?

"Well, it's not as large as I thought it would be. And I can't see the eleven cloud bands the ad promised I would see. In fact, all I can see is a faint, ruddy discoloration dividing the bright ball in half. What's wrong with this thing?"

Stay calm and collected. Remember your experience in seeing the Moon's detail. Before you start fearing you've bought a lemon, force yourself to take a longer, closer, more sensitive look-see.

"Okay."

Keep observing. Jupiter will tease you. Wait for that moment when the atmosphere stops bubbling and old Jupiter will put on a different face. Be patient.

"Yes! I just saw something. For just a second, it was banded — just like in the observatory photos!"

You feel better, don't you? That means your telescope delivers when it has a chance. And that means you will see those bands again. And when you do see them again, you will be ready to take advantage of the time and freeze that image in your mind. This image memory will stay with you and, with repeated observations, you will build up a composite of similar image memories that will be more than the sum of the parts. To do so, you must make hay while the planet shines, and learn to make seconds count like minutes.

Observing faint nebulae and galaxies is not the same color of horse. Jupiter is easy to locate in your finder. So are Saturn, Mars, Venus, the Moon, and the Sun (although doing so with old Sol without the proper filters can blind you real quick). Most nebulae and galaxies are exceptionally faint and not so easy to locate in a finder. The Great Nebula in Orion, M-42, is one of the few exceptions.

No doubt, the Orion Nebula gets great press coverage. No doubt, you've seen about ten thousand photos of it. No doubt, you are confident you know what it looks like.

What you will see when you first look at it through your telescope will look nothing like those photos. In fact, what you see just might disappoint you, especially if the sky transparency is poor or if nearby city lights pour too much cream into the universe's black coffee.

Be aware that M-42 is one of the brightest examples of diffuse nebulae in our sky. In other words, this is just about one of the easiest fish in the lake to catch. Still, you must learn how to "catch" it, and you must not expect to set the hook after the first cast. Orion may be easy, but it is not suicidal.

For me, the eyepiece view of the Orion Nebula is much better than a photograph of it. Long exposures burn out most of the subtle, soft details that contribute to its beauty. In a photo you seldom get to see the four bright "baby" stars that play together in this stellar nursery that form the Trapezium. A photo could never capture all that's going on in that starfog.

To really see the Orion Nebula, go to your refrigerator and get an egg.

"An egg?"

Yes. And some paper and a soft pencil.

"Are you sure?"

Certainly. Just trust me and another amateur named R.C. Proctor. After reading one of my articles in ASTRONOMY magazine, he wrote me a letter which made a very good point about learning to see. Believe it or not, a professional artist helped him to survive by suggesting that he exercise his powers of observation by drawing an egg. In his words:

> She offered advice that, at the time, seemed bizarre and worthless. "Learn to draw an egg," she said. "Keep observing and in the meantime draw eggs that you have placed on a sheet of white paper. Keep drawing eggs until you have drawn one that actually looks like an egg." She explained that observing your subject was as difficult as drawing it, and that to observe correctly you must look into your subject and through it with a perfectly relaxed mind and thereby detect subtle nuances and detail that escape ordinary scrutiny. Well, draw eggs I did. I drew eggs by the score until slowly they actually began to look like eggs. Not coincidentally, I began to see, equally slowly, more and more planetary detail. Subsequently, I have learned a good deal more about seeing and transparency and eye fatigue. Now, when hearing a novice astronomer deride his fine new scope, my first and best advice: Draw An Egg!

Feedback like the above from survivors such as Mr. Proctor makes everything — all the long hours of observing and all the times my tired brain has hurt from lugging sluggish words onto paper — worthwhile.

Teach Your Eyes to See. If you hate chickens, you don't have to use an egg. Try a pencil. *Stare* at it. See a pencil like you've never seen one before. Notice the imperfections in its paint job and the texture of its eraser. Why does the eraser look soft and the pencil hard? Something about the way things look tells you such things — but what?

Better yet, stare at a blank sheet of typewriter paper for hours on end. I do this a lot, being a victim of frequent writer's block. Most people would describe a sheet of typewriter paper as featureless white. An astronomer could tell you right off that this isn't the case. It is less dense in some areas, more dense in others. It is textured — very much so. Study it closely and you will detect the cotton fibers. Study it even more closely and it will take on the characteristics of a nebula, with sworls and eddies. Every sheet of typewriter paper has its own unique complexion, if you take the time to learn to see it.

Once you're learned to see, you can then take advantage of certain accessories that will help your trained eye to see even more. I'm thinking of relatively inexpensive nebular filters (which block natural skyglow and sodium and mercury emissions from those confounded vapor lights) and photo-visual color filters (make sure these are made from top quality optical glass and will thread directly into your make of eyepieces). This sort of help cuts out all the static on your optical radio — but you won't notice the difference if you're not aware of the static in the first place.

* * *

Now, as to what you can do to control the other type of seeing:

Each and every day our atmosphere gets more polluted. You know that. Every time you start your car you add to the filth that is slowly but surely blotting out the stars. Every time you see some factory pouring junk into the air and don't try to do something about it, you are making it rougher for your telescope to please your eye.

We are amateur astronomers who profess our love for observing the night sky. Are we going to sit on our tails and let uncaring people continue to soil our window to the stars? Our atmosphere is already showing the first symptoms of what could be a terminal illness. To a non-astronomer living away from smog-laden cities, the sky is still blue, the night sky still abounds with stars, and the sunsets are even more beautiful than ever. They are not even aware that such gorgeous sunsets are, in fact, symptoms of the atmosphere's disease. The more beautiful they get, the sicker our air is getting. Only someone who knows the sky and the effects pollution has over even unpopulated areas can get a feel for the seriousness of what is happening. Humanity is good at shrugging off things until they strike close to home. If you are a dedicated amateur astronomer, air and light pollution is trying its damnedest to destroy what you love — our cosmic connection.

Don't let pollution strike the final blow. Write your congressman and demand that he fight for even tougher air-quality controls. Ask him to do something about all the needless light pollution. And get tough with him. Tell him you won't vote for him unless he does.

Let clear skies be our legacy to the astronomy enthusiasts of the future.

10 Observing in the Wilds

SO YOU'VE BOUGHT A TELESCOPE, become familiar with its workings, and have begun to learn to see. Now, let's walk a little farther into the jungle and talk about actual observing sessions — those times when you really get down and boogie.

If you live in or near the city lights and refuse to quit your job and homestead for some outland up in Canada, most of your observing will be conducted at home, under a night sky that is quite a few shades short of being dark. For a true astronomer, living in or near the city lights is like living at least thirty miles east of Eden. It may not be perfect, but it *is* home. And you must make the most of it. After all, it is unrealistic to think your health could hold up to trucking your scope out to dark skies every clear night and then lugging it back in time to get to work, asleep on your feet.

Lovers of lunar, solar, planetary, and double-star work do not suffer so much in this type of environment. Lovers of deep-sky work find it a living hell. Their nebular filters ease the frustration in some instances — but still, most of their time is spent agonizing for darker skies.

There's no doubt about it: Observing is best when done in the wilds. I don't care how long it's been since you mowed your backyard, or how mean your neighbor's dog is, or how big the mosquitoes are around your house — setting up your scope at home and observing there is not observing in the wilds.

I'm talking about the real thing, folks: where tumbleweeds tumble; where the deer and the antelope play; where wolf packs roam in search of fresh meat; where skunks sneak around; where snakes slither; where creepy-crawly things come out from under dead logs to prove their macho in the night. I'm talking places where darkness dwells and shadows take on hideous dimensions; places far removed from 7-11s, and restrooms, and vapor lights, and civilization; areas where the dark forces commune to utter all sorts of weird groanings; places that animals you only see on Mutual of Omaha's *Wild Kingdom* frequent. The carrion-eating birds' cafeteria. The bobcat's playground. The bear's domain. The coyote's kingdom. The armadillo's stomping grounds. Where skunks hang out.

Observing at home, you may encounter problems (power failures, nosy neighbors, unexpected company, late-night tax audits, nosy policemen, nosy newspaper boys, passionate alley cats, or smoke from your neighbor's outdoor grill), but none that can kill you or maim you or scare you so badly you wish you were dead. And at home, the bathroom is very near.

Unless you live in the boonies, observing at home is nothing close to observing in the wilds away from the skyglow of city lights. In the wilds you can see so much more even with the naked eye. Even if you can only set aside a few nights a year for remote observing, do it. Take the time to find a good, remote site preferably at least thirty miles from the nearest city. If it is not public land, get permission to use it. It is my experience that most land owners are very receptive to amateur astronomers. After all, you're not out to kill their game or harm their land. Be sure to invite them to one of your observing sessions. They probably will not take you up on the invitation, but they will appreciate the gesture. Always remember to clean up your observing site when you leave.

When you consider a site, check first for some level ground relatively free of obstacles you could stumble over and break your neck on in the dark. Pick a spot that is easily accessible for your car. If a large city's skyline is visible — even from a distance of thirty miles — pick a clearing with trees blocking the horizon in the direction of the city.

The ideal spot is a clearing about a hundred feet in diameter surrounded with relatively short trees. Mountain tops are usually more trouble than they're worth.

Decide on the best spot to set up your telescope and pick an area at least twenty feet in diameter clean of all large rocks, sticks, and cow patties. The area immediately surrounding your telescope should be scoured clean of debris of all sizes. Even pebbles can be rough on the knees, believe me. And make sure there are no ant beds nearby.

Next, you should give some thought to your restroom. This is a touchy subject, to be sure — but something you must deal with unless you don't do that sort of thing. Of course, in the wilds a restroom can be anywhere, as long as it's a respectable distance from your observing area and not near the bush where the skunk lives. So, before you set up for observing and while it is daylight, pick the area for your restroom and give it the once-over. Police the area. This will also spread your scent around to discourage creatures and make them shy away. You may think all of this is silly and useless. You will not think so the minute the urge hits you and you are forced to seek relief in some clump of brush unfamiliar to you. Obviously, men have it easier, but ladies, you know and I know that you are especially vulnerable at such times. Take the necessary precautions, please.

Once you have selected and made ready a remote observing site, you are ready to get everything in order to "boogie." Preparation for this begins on the home front, long before you go out to observe.

If it happens to be winter, I shouldn't have to tell you to dress for the occasion. Denim jeans are not the best insulators unless you wear good long johns under them. The ideal material is wool, not cotton. Insulated socks and insulated footwear are a must. If they're high-top and snake-proof, all the better. Dress so your skin can breath, so the perspiration (yes, you still perspire in the winter) can evaporate and not collect to rob your body of warmth. Since you will be working a lot with your gloves off, purchase a hand warmer; they are invaluable when it comes to fighting off finger numbness caused by cold. Chances are your nose will run — and so will your eyes. Take along a clean handkerchief. Do not dress bulky. Bulk will not keep you warm, and it will make it all the more difficult to get into the 734 weird positions you will, at one time or another, have to get into to observe.

If you wear glasses, purchase an extra pair and keep them in your car. Many amateurs who wear glasses do not like to leave them on while looking through the eyepiece — but most have them on when sighting for a new object. Weird things can happen to glasses during these times of on-again, off-again, on-again, off-again. They can end up underfoot, crushed flat. Without the extra pair in the car, you would have to drive back home half-blind.

Before you go to observe in the wilds, make a check list of the items you wish to take. Don't trust your memory! I learned my lesson the very first time I drove fifty miles to a remote site only to find that I had left my eyepieces on the kitchen table right next to my toilet paper.

Sample Checklist

☐ Telescope
☐ Eyepieces
☐ Star chart
☐ Red-filtered flashlight with extra batteries
☐ Coffee (or whatever beverage you choose to imbibe)
☐ Snack
☐ Allen wrenches, screwdrivers, all other necessary tools
☐ Camera
☐ Film
☐ Observing log and two sharpened pencils
☐ Aspirins or non-aspirin pain reliever, allergy pills, needed medicines, etc.
☐ Toilet paper
☐ A first-aid kit
☐ Bug repellant
☐ Raincoat
☐ A baseball bat (This little bit of protection will make you feel better.)
☐ Handwarmer (in winter, of course)
☐ Eyedrops
☐ An accurate watch or clock set to the correct time
☐ Detailed, step-by-step observing plan for that specific night

Do not check off this list in your garage or kitchen. Check each item off as you put it into your vehicle.

The detailed observing plan is very important. Using the month's star charts in either ASTRONOMY or *Sky & Telescope* magazine, and taking the suggestions included in the observer's pages of each magazine, list each object and memorize the location of each object on the star chart. Since you have done your homework and are familiar with the constellations visible at this time, you will immediately go to work and not just wonder what in the heck you are going to look at. Each minute of observing time at a remote site is precious, especially if the seeing conditions are perfect. You certainly don't want to waste time at the ball game wondering what you want from the concession stand.

Before you drive off, leave a note on your kitchen table telling where you will be. Sketch a map if you have to, so should you be needed or should something — God forbid! — happen to you, the right people will be able to locate you easily. If you live alone, it is a good idea to tell someone close to you just where you will be and when.

Make sure there is plenty of gas in your car and that your spare tire is okay. Leave early enough to get to your observing site before sundown, if possible. Better yet, get there in plenty of time to completely set up (minus the aligning, of course) while it is still light. Get everything ready so that when you get the first glimpse of the pole star, all you have to do is align the scope, nothing more.

Do not build a campfire! All that will do is fill the air with a low-lying cloud of smoke. You came out there for clear skies, remember?

***

It's when the Sun goes down and the dark sets in that, if you're not careful, mother nature will test your marbles with her horror show. If you have a friend with you, things will not be so spooky. But if you happen to be alone, your private little site can get haunted real fast. You will soon develop the most acute hearing of your life. A snail crawling two hundred feet away will sound like the Blob. Trees will shed their limbs in such a fashion that you will be positive that someone is sneaking up on you. Invisible (and sometimes visible) eyes will stare at you from the treeline. You will forget to swallow. Blood will roar in your ears and your chest will hurt because you haven't taken a breath for three minutes.

You may even hear a snort.

If that happens, in two seconds you will be locked in your car and panting like a moose at bay. You will switch on the car's headlights to see your attacker. It may be an armadillo rooting for food. If so, the headlights won't even phase it. You see, the armadillo is too dumb to be scared. The armadillo is too ugly to be scared. And too, he doesn't believe in things that go bump in the night.

You don't either. Not when you are safe at home at night you don't — but out here in the wilds, it's different. Out here, things *do* go bump in the night. Dead leaves crashing to the ground go bump in the night. Night crawler worms colliding with tree roots go bump in the night. The heartbeats of bunny rabbits go bump in the night. Droppings from snoozing birds go bump in the night. And God help you if there happens to be a screech owl nearby.

Daniel Boone-type or not, chances are your first night of observing in the wilds will be a miserable one full of stark terror. Don't be ashamed or embarrassed. You're not a 'fraidy cat. Stick with it. Sweat that first night out. Gradually, these discomforting feelings will disappear. It'll be like living near a train track. Later, the only thing that will wake you will be a sudden train strike. When the 11:15 doesn't rumble through on time, you will wake and say, "What was that?"

You will get to the point where all those strange bumps in the night will go away. The shadows will stop moving. Certainly, there are animals in the wild that can hurt you — but for the most part, they don't want anything to do with you. Contrary to popular belief, it is not easy to get eaten by a wolf pack. Wolves spend most of their lives getting shot at by two-legged creatures like you. Unless you are already half-dead, they'll steer clear of you. If you are observing in cougar country, make plenty of noise. Take along a transistor radio and let it play. Mention out loud that you have that baseball bat I told you to take. Cougars know it is much easier to catch a rabbit for supper than a human being. And moreover, rabbits don't come pre-basted in insect repellant.

I have observed in cougar and wolf country for years, and I have never seen either animal. The only critter that has ever scared the pants off me was a skunk. You will come across them from time to time, and if you observe in the Southwest, armadillos. Bear country requires taking more precautions, especially if you observe in a national park where the bears are less afraid of human beings. In that case, talk with one of the game wardens and get the nitty-gritty about bears.

Here in Texas, my definition of a truly dedicated observer is one who can successfully complete a hand-guided 42-minute exposure of the Dumbbell Nebula with rattlesnakes coiled around his ankles and mud-dauber wasps building nests in both ears. Well, I don't qualify. If I even smell a rattlesnake, I pack up and go home. There's no sense getting too serious about this astronomy business!

Observing in the wilds with friends can be very enjoyable, providing each friend has his or her own telescope. Let's face it, sharing a scope is not the cat's meow. There is nothing more tedious or nerve frazzling than to stand by while another gazes into your telescope and ooooooooooohhhhhhs and aaaaaahhhhhhhhhs for what seems like two hours. Your friend will suffer the same when your eye is stuck to his eyepiece. If your friend cannot go the expense of buying his own telescope, see if he will invest in a pair of astronomical binoculars. That way, when one is observing at Big Daddy, the other can be enjoying the low-power, wide-field views through the binoculars.

And speaking of binoculars, forget the cheap, department store models of these also. I don't care how crisp and clear the parking lot of the store appears when you test them out from the counter. Cheap binocular lenses will turn stars into seagulls. Purchase yours — or talk your friend into getting his — from a reputable telescope dealer who specializes in binoculars suited for astronomical applications.

* * *

If you can join an active club, chances are they will already have a suitable, remote observing site. If you're really lucky, the place might even have restroom facilities, electrical power outlets, and concrete pads for your telescope. Even so, take along a roll of toilet paper, just in case.

11 What Should You Specialize In?

NOT LONG AFTER YOU GET THE kinks (if any) worked out of your new telescope and accessories (and perhaps, back), you will probably grow a bit disenchanted with casual observing and feel the need to specialize in one certain area.

"What about the Moon? You said that that was your choice."

Yes. I chose the Moon because I enjoy making pencil sketches of the Moon's surface features as they appear in medium- to high-power eyepieces. For nearly twenty years I have taken great pains to do this accurately and artistically. I'm not so naive as to think my sketches will contribute anything to the bulk of lunar knowledge already stashed away in the vault of science, because they won't. I'm sorry, but I don't sketch the Moon for the benefit of science. I do it for me. Certainly, sketching old Luna is not as important as making long period studies of the Moon's albedo changes or searching for transient lunar phenomena — but for me it's not as boring either. It's fun. I enjoy it. Sketching the Moon is not something I have to force myself to do.

Granted, the Moon has a lot going for it. It's easy to locate. It's up practically all the time. It has a marvelous, hideous, desolate, beautiful complexion that lends itself to easy study through amateurs' telescopes. But just looking at it time after time can get boring, believe it or not. The chances are that you will not see any changes in its surface detail throughout your entire observing career. You will read about obscurations, flashes, and colored glows, but don't nurture any hopes of witnessing such phenomena. You have a better chance of winning a million buck lottery.

If geology doesn't interest you, I would suggest that you choose another specialty area.

If you are intent on contributing something useful to science, realize that such opportunities in the lunar field are almost nonexistent where amateur work is concerned. The few that do exist involve colorimetric work, timing occultations, studying eclipses, and photometry: all of which require a great deal of effort, dedication, patience, and skill.

True, colorimetric work involving the Moon sounds impressive. Just don't choose to tackle it to impress your astronomer peers. Leave that sort of thing to the weirdos who specialize in things like searching for gas giants rich in Corning Ware or studying the effects of airborne pollen on the diffraction rings of fourth magnitude stars. Tackle it only if you know you will have fun trying to tackle it. The same goes for sketching, occultation timing, eclipse study, photometry, or astrophotography.

Make sure it will be fun, then go ahead.

"The planets have always interested me."

The planets interest every amateur astronomer to some degree — but unless your telescope has a larger-than-average objective, you will have to be content with focusing on Mars, Venus, Saturn, or Jupiter.

Mars

As I warned you earlier, Mars will probably burst your bubble. Even at opposition, its disk appears small. Its surface details are of low contrast most of the time. You will need a quality long-focus refractor or reflector to get the most out of what little Mars offers, and you will have to use high powers at that. If the seeing is not exceptional, there's no sense in even considering the planet. Usually, it's not worth the time or the trouble, especially if you're a tenderfoot — and you *are* a tenderfoot.

Venus

It is bright. It does undergo phases like the Moon. Other than that, just about the only other thing you can say about it is: You need a good pair of Foster Grants to look at it. Venus' clouds get a big kick out of throwing the Sun's light back at you — so much light, in fact, that you will need filters to deal with it (unless daytime or twilight viewing is your bag). But even with the proper filters, precious little detail can be observed. If you think you might like ultraviolet photography, Venus is the lady to start out on.

Saturn

Saturn will not disappoint you. I don't think anyone believes it is really up there until they see it with their own eyes. It is truly one of the most awe-inspiring objects in the heavens. The rings *do* exist, and you *will* be able to see them — not just think you see them. You won't see many cloud bands, but you won't care. You may not even see Cassini's division of the rings — but, again, you won't care.

There's a lot of fun to be had observing Saturn, so be my guest!

Jupiter

The big guy. The one with the marching bands.

Who wouldn't have fun watching Jupiter's four brightest moons play tag around Big Daddy? Who wouldn't have fun watching all those cloud bands perform?

Jupiter, even in moderate seeing, gives a good show for the trained eye. Its disk is big enough that it will provide pleasing views in even the smaller sizes of amateur telescopes.

Try it; you'll like it.

"Maybe I will — but what about double-star work?"

It doesn't turn me on, but it does turn on a lot of amateurs and professionals.

For profitable double-star work, you must have a telescope of long focal length and exceptional resolving power. You will also need much patience and discipline, as your observing must span years and be adorned with exact records and measurements. I can't for the life of me see how anyone could consider observing double stars fun — but one man's junk is another man's treasure. I use double stars (binaries) for the sole purpose of testing my telescope's performance.

"What about comet hunting?"

Not a bad choice if you can hack hunting year after year after year without discovering one. If you've got an easily-handled telescope of short focal length that provides good wide-angle, low-power views, then this might just be the specialty field for you. Just don't count on discovering a comet. Rumor has it that the Japanese bought that franchise long ago, and there is some truth in that. The Japanese have cultivated persistence, patience, and dedication for thousands of years. Japanese comet hunters are prime examples of the resulting harvest.

No other work in the field gives the amateur such a chance at virtual immortality. Think of it: A comet with your name!

If you think comet hunting will be fun, give it a try. Just be aware that many, many comet hunters have searched diligently for years without discovering a new one.

"There's always deep-sky work."

Deep-sky work can be a wild and woolly trip from one end of the universe to the other. It can mean trying to observe every object listed in Messier's catalog . . . and then some. It can mean studying the Orion Nebula for hours on end, seeking to see more each minute, and seeing more each minute. It can mean observing stars so distant that they merge into balls of starfog. It always means testing the limits of your visual acuity.

"What about variable star observing?"

It will certainly keep you busy, though it's not the most exciting thing I can imagine.

"And meteor observing?"

Meteor watchers pursue their hobby while lounging in comfortable chairs while their telescopes are safely locked in their storage closets. It can be fun, but why specialize in it? You've just spend a wad of money on a telescope.

"Why do I have to make a decision?"

You don't! That's the surprise. You can have fun with all of these areas and more. Chances are you will gravitate to a specialty interest. Just don't force it. Let it happen. Let the universe direct you, not vice versa.

That's the only fun way.

12 Blunders and Bloopers and Murphy's Law

"If anything can go wrong, it will." — Murphy's Law

I DROPPED A BEAUTIFUL NEW 12½-inch f/6 mirror once.

Murphy made me do it. Either that, or he made the mirror do it.

I can't even begin to describe what the concrete floor of my garage did to that glass or my nerves. The word "shattered" isn't nearly strong enough to use here. Little shards of that long-lost mirror still sparkle at me from time to time from deep within the old floor's cracks, each one a grim reminder of my butterfingers.

Remember? I told you glass was unforgiving!

Needless to say, after that champion blunder happened, I sort of died inside. I stormed into my house vowing never, *ever* to bother with astronomy again. My wife said: "You wrecked the car last summer. That didn't stop you from driving. It just made you drive more carefully."

Maybe you're talented. Maybe you're extremely careful. Maybe everyone interested in astronomy and living within a thousand-mile radius of your observatory would trade their souls for your equipment. Murphy don't care!

You — yes, YOU! — are going to suffer your share of blunders and bloopers, too. You might as well accept the fact. It is as certain as death and you-know-what. The only thing that will help you to survive such tests will be (1) the misery-loves-company awareness that you are not the Lone Ranger, and (2) a healthy sense of humor.

During my tenderfoot days of astrophotography, blunders and bloopers were a dime a dozen. For a long while, I expected the night's exposures to qualify for federal disaster aid. I still have boxes full of negatives and prints covered with out-of-focus smudges, nervous star-trails, and over- and under-exposed images of the Moon and planets. Outstanding is one unprinted roll of 35mm Plus-X film. Each frame contains two overlapping images of the Moon. Know what I did? I exposed the roll, rewound it, removed it from my camera, placed the roll into my film box, took a much-deserved coffee break, returned to my film box, proceeded to reload my camera with the same roll of film, and then took 36 more exposures of the Moon. Wanna know something else? Close examination of the developed film shows that over half of those darned exposures would have really been great shots had I not stacked them one on the other. In my log book, I have that night's seeing described as ". . . one of the five best nights I've ever encountered!"

Astrophotography gives old Murphy so much to work with. Just bagging the game (taking the pictures) is not enough. You must then dress and prepare the meat (develop and print the pictures), or get someone else to do it for you. Here is where Murphy can turn you (or that someone else) into a real butcher, if you get my drift.

If you choose to place your film at the mercy of one of those quick developing and printing services, don't expect preferential treatment. Your celestial masterpieces will be run through the mill with everyone else's vacation shots and baby pictures. And your precious roll will probably be only the second roll of film the company's machine has chewed up in nearly eight months, which is a coincidence since the *first* roll happened to be yours also. The free roll of film they might give you will be of little comfort.

WARNING! Using a fast-film service to process and print your astrophotos is not advised. There's no sense in making Murphy's hobby too easy for him.

Custom film processing and printing is not advised either. An expert in developing and printing lovely portraits or scenic shots may not know a hill of beans about developing and printing a long exposure of the Orion Nebula.

The best route is to learn to develop and print the buggers yourself.

"But that means buying the chemicals and equipment and managing some sort of a darkroom setup."

Hey! Go to the head of the class. You're correct. And that also means more financial outlay and more hard work and more things Murphy can sink his teeth into.

"Maybe I won't give Murphy a chance in that department. Not

for a long time, at least. Maybe I'll just observe."

That's up to you, but it won't make you off-limits to Murphy. He can and will pick on you when you least expect it.

I lectured once at a junior college on observing Jupiter. I walked into that classroom prepared and confident, armed to the teeth with an impressive battery of graphic aids. I was really truckin' when a student lifted his hand and asked this question: "Sir, why do you keep calling Jupiter *Saturn*?" Needless to say, my credibility took a very steep nose dive. Murphy's a specialist at low blows.

I was interviewed live on camera by our local boob-tube station only a few minutes before a total eclipse of the Moon. Remember, I'm a Moon specialist. I was prepared and confident when I addressed that reporter's mike. Right off, he asked me the diameter of the Sun! I couldn't think of it to save my very reddened face. Murphy had struck again!

I gave a speech at a ladies' club luncheon and bored most of them to tears. They'd come expecting to hear an intriguing talk on astrology, not astronomy. I should have smelled a rat right off. I'll give you two guesses (first guess doesn't count) just what the program chairman's last name was. You guessed it. Murphy!

I mailed an article entitled "Housewives Can Be Astronomers, Too!" to a slick ladies' magazine and got back the manuscript with a short, handwritten rejection note saying, "Would you rewrite this using the astrology angle?" Astrology indeed! Murphy was just setting me up.

I used to think I was impressing people by driving around with my 12½-inch plywood Dobsonian until I heard that one of my neighbors wouldn't let his kids come over and play with mine because I drove around town with a coffin sticking out of the back of my Mazda.

Murphy never tires of pulling the same old tricks. So many times I have tripped on my telescope's power cord and jerked the mount out of its carefully aligned position. I've stumbled over the tripod legs at least a thousand times — but I'll take responsibility for not more than three hundred. No matter how careful I am, I bang my Dobsonian tube assembly on something, knocking the optics out of collimation with disgusting regularity. Murphy causes me to drop eyepieces, step on eyepieces he won't let me locate in the dark, and breathe my own dew on the crazy things more times than I care to note. He even makes me forget things I've checked off as loaded on my checklist. After nearly twenty-five years, I'm still the King of Blunders and Bloopers.

Sure, there are still times when I consider turning my attentions to a much less demanding hobby, like rubber band collecting. That doesn't last for long, however. Can you imagine what Murphy could do with something as unpredictable as rubber bands? The mind boggles!

Blunders and bloopers will frustrate you. Sometimes, they will embarrass you. But as long as you know that you are doing the best you possibly can, roll with the flow and keep your sense of humor. Never give up. Thinking about giving up is okay, even therapeutic. But as soon as you admit defeat, you're done for. The universe will deem you unworthy and turn her back on you.

Murphy will tag along for the duration of your enthusiasm — but never make it easy for him to sting you. Each step you take into the jungle, go slowly. Plan every move. Ask yourself what could possibly go wrong if you step here or turn this way or that? Think as you move. Use some common sense. Get tough! Keep on keeping on. If Murphy sees that he's not making you miserable, he just might turn his attention elsewhere for a while. You must be ready to take advantage of these times.

And too, sometimes Murphy will cut his own throat, and two wrongs will make a right. Once I purchased a very expensive wide-angle eyepiece when I could not afford it — Wrong #1. I eventually sold it for a loss only a month before the market price for the things almost doubled — Wrong #2. Disgusted with myself, I took part of the money and purchased a less expensive Erfle only to find that it actually delivered much better views! This eyepiece is still my favorite eyepiece. Chances are I would never have bought it if Murphy hadn't gotten overzealous.

I'm tough now. I can laugh — even while crying — at my blunders and bloopers. It can be likewise for you, if you survive.

Get mad and kick and spit — but don't quit.

And don't forget to laugh.

13 Living with a Non-Astronomer Spouse

IF YOU ARE MARRIED, AND IF YOUR spouse could care less about what makes Saturn run, there is a definite possibility that your telescope could become second party in a divorce suit. And Heaven help you if your spouse's lawyer decides to stick you with f/8 null-figured adultery!

When you love something so much, it is very easy to neglect the other loved ones in your life. It is very easy to resent anyone who does not share your passion. This is especially true where amateur astronomy is concerned. I, for one, cannot imagine why anyone would not be in awe of the Dumbbell Nebula, or the Double Cluster in Perseus, or the Ring Nebula in Lyra, or the crater Plato. And it was some time before I got over the urge to call anyone who did not share these feelings a stupid, blithering, insensitive, idiotic, half-blind, ignorant nerd!

My sweet wife tried her best to be interested — and she is . . . in a way — but she will never be an amateur astronomer. And it's not because she's a blithering nerd. On the contrary, she's a brilliant woman. She's a very talented artist who regularly sells her paintings.

I'm not a painter. I would like to be able to paint as well as she does, but I haven't the talent. That special door was not opened to me. I have not been called to be an artist of the canvas. I appreciate good art, but I do not harbor the desire to walk in the artist's moccasins. I have no right to expect my wife to embrace astronomy; she has no right to expect me to embrace art. What's more, we don't expect it of each other, and our marriage is rock-solid (boulder-type rock-solid). We meet each other halfway. There have been nights she has wanted to go out — clear, perfect nights for observing — when she has, instead, offered to stay home with the kids while I took advantage of the good seeing. In turn, there have been perfect nights when I have insisted on going out to a movie. Like drinking, the key is moderation.

Little things can ruin a good marriage for an amateur astronomer. Little things like saying:

"I've just sold our roof to get a dome."

"I can't believe we've got twenty-seven scopes on lay-a-way!"

"So what if the power company shut off our electricity? It'll make for darker skies!"

"Honey, I've just clinched a deal on 643 fiber glass telescope tubes, so you're gonna have to get two jobs."

"Sweetie, I know we've never had a vacation, so how'd you like to take a nice tour to the Heart of Darkest Africa, sleep with some neat headhunters, stare at the Sun, and then be in some pretty heavy shade for about two minutes?"

Big things can ruin an astronomer's marriage: Total lunar eclipses on the night of your honeymoon. Selling the family car to make a down payment on a used mountain top. Filling your deep freeze with film instead of food. Giving your wife a variable frequency drive corrector for your anniversary, when she doesn't know what it is and doesn't care. Whispering the name of your telescope in your sleep. Removing all the pictures in your wedding album and replacing them with high-resolution shots of your favorite lunar craters. Waking up late at night after hearing a suspicious noise in the den and asking your wife to check it out while you sneak out to the garage to protect your telescope. Replacing your spouse's photo in your billfold with a photo of you-know-what. Hocking your wedding ring to finance a new focuser. Allowing ten observing buddies to move in with you. Spending every extra cent you have on your telescope, or accessories, or more catalogs to get you hungry for more. Deciding to begin a life of crime to finance your next telescope purchase. And so it goes . . .

Seriously, though, marital frictions can arise from eating, drinking, and sleeping amateur astronomy. It can also reap havoc with your social life if you happen to be single, not to mention your job if you fall asleep on it because you had no sleep the night before.

Budget your time and your money and stick to it. Be gentle and understanding with those who do not share your enthusiasm. Spend more time with your loved ones than you spend with your telescope. Be sure that they know *they* come first in your life. If you follow these guidelines, your wife or your husband will go to greater lengths to make the time available for you to pursue astronomy guilt-free.

There have been many times when my wife has stepped outside to check the sky. It is always a special gift for me when she comes up and says: "It's a nice night. Why don't you take advantage of it? I've fixed you a snack and some good coffee. Have fun."

14 Is the Universe a Harsh Mistress?

WE'VE GONE PRETTY DEEP INTO astronomy's jungle. I feel we are pretty close now. May I burden you with a personal problem? Something is really bothering me.

As you know, I'm tough. It takes something pretty darn shocking to bother me. What's bothering me is tittle-tattle.

You see, there's this rumor going round that the universe is a harsh mistress, that she seduces fledgling astronomy enthusiasts into virtual bankruptcy with her beauty, and then proceeds to chew them up and spit them out much the worse for wear.

I know firsthand that this pitiful crumb of idle gossip has already gathered a savage harvest in eleven American states, England, and Australia. In each of these sad cases, the rumor proved to be a grim reaper of sorts whose scythe severed a person from his or her relationship with the universe, leaving that person bitter, disappointed, confused, and angry.

How do I know this?

Letters, folks. And telephone calls.

Because I've written well-read and -received articles for ASTRONOMY magazine concerning the ups and downs, dos and don'ts of amateur astronomy, I have received and continue to receive letters and phone calls from people all over the world. In some cases, this reader feedback was written to me in care of the magazine and forwarded to me by the kind editor, Richard Berry. These — including those letters to the editor which were published in ASTRONOMY's letters column — happen to be very complimentary, the stuff of writer's dreams. They made me feel good because, obviously, my articles had done for the readers what I had intended them to do: help make the way a bit less bumpy and dangerous for potential amateurs just beginning their trip into astronomy's jungle.

Quite honestly, I was very high on myself for a good while. Later on, flies started turning up in the ointment. I began receiving distressing epistles directly in the mail. These letters scraped living tissue from my astronomer's soul. But the phone calls cut even deeper; they were cries of anger from disenchanted people who were standing out on a ledge, threatening to jump and end it all.

No, these were not cases of threatened suicide. They were individuals calling or writing to say something like this:

"Hey, Mr. Know-It-All. I've really tried to get into this astronomy thing, and I just can't hack it. It's not what I thought it would be. It's not what *you* promised it would be. I want to get out!" Either that or they wanted to sell me their telescope equipment for cheap — in most cases, dirt cheap.

"What's wrong?" I asked each one of them.

Not one of them could answer that question. No one could put his or her finger on the source of the anger. But the more I talked with them, the more I realized that all had several upsetting things in common: they'd gotten interested in seriously pursuing amateur astronomy after reading one or more of my articles; they'd invested a great deal of money in the best astronomical hardware available; they'd purchased the most advanced astronomy books and the most expensive star atlases; and now, they felt cheated.

Since I was the spider who had invited them into the parlor, they held me accountable. They blamed me! I don't mind telling you that it was a long, long fall from a high, high mountain top for this stargazer/writer. When I hit bottom, it hurt. After the fourth call and fifth letter, I figured out what had happened to these people. I was able to pick myself up, dust myself off, and doctor my wounds. I felt better about myself, and a little ashamed for having let myself fall in the first place.

What had happened?

First, not one of them had understood the real message word-woven within the tapestry of my articles. Second, the rumor I mentioned earlier had gravitated to them, and had done its dirty work. You must understand that this tittle-tattle ferments the sour grapes of certain people who have tried and failed and been found unworthy of amateur astronomy. The wine of their misery is bitter, and they are always willing to share it. If you are weak and unworthy, one sip of it will dissolve any relationship you have with astronomy — but not before you, too, pass the cup. Misery loves company.

I gave these people the benefit of the doubt. I re-read my articles, three of which are chapters in this book. Yes, each one said exactly what I had wanted it to say. The message was loud and clear. I was satisfied that I had not supplied the evil potion that had turned these people mad. Certainly, they'd drunk the wine of sour grapes — but I had not served them the drink.

What, I wondered, had driven them to drink in the first place? After a little thinking, the answer was painfully obvious: in each case, they'd made the mistake of thinking they could buy the favor of the universe with cash instead of love and devotion and hard work. All of them had OD'd on astronomy hard- and software before paying their dues and getting in shape. They'd started off taking their first "baby steps" weak from a siege of the fever picked up from reading the ads, various books, and articles (mine included) that glorify amateur astronomy. If you're really into amateur astronomy, you know the fever. It doesn't make you sick at all; on the contrary, it is a very rich, albeit nervous, high. It can impair your judgement. If you're not careful, you may not be responsible for your actions.

Some people can deal with the fever. Those who *do* usually go on to become successful amateur, and sometimes professional, astronomers. Those who fail sometimes come out of the fever screaming, angry, and bitter — easy prey for that vicious rumor beastie.

The really weak ones crash. Over the phone, I heard their crashing. If you love astronomy as I do, there is no sadder sound. Wanna talk about feeling helpless?

I'm not a psychologist. What could I tell them? What magic words might I have said that would have pierced their drunken stupors and dissolved their disillusionment and bitterness? What lifeline might I have thrown them to save them from the quagmire of self-pity?

They told me horror stories: they said their telescopes were a bunch of junk, would not show lunar craters clearly, or the canals (?!) of Mars, or the spiral arms of galaxies, or the colors of the Orion Nebula. One man, especially, picked on the Orion Nebula, and during the entire long-distance conversation, referred to it as M-31. I didn't have the heart to tell him that M-31 is not the Orion Nebula, but the Andromeda Galaxy.

I knew what I wanted to say, but how do you slap some stranger in the face with the cold fact that it takes a great deal of time, effort, and skill to see and appreciate the subtleties of nebulae and galaxies and planetary detail? These people expect — no, they demand — that each and every object appear in their telescope like it appears on a Palomar print.

How does one kindly inform someone that they haven't earned the right to see the heavenly wonders?

One gentleman complained that he had wasted well over two thousand dollars on a 3-inch refractor, observed with it one time, and found that he could not discern more than two cloud bands on Jupiter (and no surface detail on any of its moons!). In utter frustration, he wondered if I'd take the pile of junk off his hands for half the price he'd paid for it.

I was sorely tempted. His refractor was one of the best, made by a company with an impeccable, long-standing reputation for delivering the finest state-of-the-art optics. I knew the telescope was not at fault — he'd simply not given the telescope a decent chance. And his mind was made up; the blankety-blank scope was a piece of junk!

What could I say? At the time, three of my close friends were working day and night at two jobs each just to save enough to purchase telescopes costing half of what this man's cost. It would have been oh, so easy to take advantage of the situation without saying another word.

Would it have made a difference if I had told him that there have been many nights when I have not been able to make out one cloud band on Jupiter with my 4-inch refractor, when the Moon's craters jiggled in my eyepiece like hula dancers, or that I have never observed any surface detail on Jupiter's moons because that's impossible with an objective of such diameter?

I tried — I really tried — to explain the bad seeing, the Dawes Limit, the realistic capabilities of his fine telescope. Metaphorically, I explained that a .22 rifle, no matter how well-made or straight-shooting, cannot have the impact of a cannon. Would he listen? No way, bluejay.

One of my friends now has that telescope, and every clear night, he's out exploring the universe, happy as can be, and very grateful to me for finding him such a prize precision eye. It was, after all, the least I could do for both parties.

One man's junk is another man's treasure. The individual who considers such an instrument a treasure is worthy, and deserves to have it. The man who considers it junk will never be an amateur astronomer. Now, the man still blames me for using "pretense" to arouse his interest in astronomy, the telescope company for ripping him off, and the universe — if you can imagine it — for not delivering the goods!

Yes, Virginia, some people are unworthy.

And speaking of Virginia, I received this very angry epistle from a woman who lives in that beautiful state. She was mad also, though not at me. She couldn't figure out how in the world the company she'd bought her telescope from could have the nerve to label her 12.5-inch reflector as "transportable." She couldn't even lift the tube assembly! And for that reason, she'd never observed with it. Would I be interested in buying the monster cheap? I thought of my friends again. I couldn't help it.

Even so, I tried to cool her boiling anger. I informed her that the company (I have its catalogs) fully describes the weight and dimensions of her telescope. Should I have added that, had she done her homework before ordering, she would have realized that a tube assembly for a 12.5-inch f/6 reflector has to weigh a bit? And that it needs to be about six feet long and fifteen inches in diameter? Should I have gone on to explain that, logically, an equatorial mounting for such a tube assembly would have to be of sufficient mass to support the whole works properly? In such cases "lightweight" means you don't need a crane, but you may have to strain.

Perhaps I shouldn't have, but I did.

Well, she didn't like it one bit. She called me arrogant. I felt for her, but I didn't feel sorry for her. Increasingly, I find it difficult to feel sorry for anyone who does not use any common sense. Needless to say, I found her telescope a loving home.

Common sense is important, but it isn't the whole ball of wax, either. Consider the following caller: He was angry that it was taking far too long to get his telescope. When he ordered, the company informed him in writing that delivery would be made in approximately four months. When the man called me, the telescope was six months overdue. He was absolutely livid.

What could I say? The same thing has happened to me on three — count 'em, three — occasions, with three different companies. I knew what the man was feeling, but I also knew that there was no magic method known to me, short of marrying the boss's daughter, that would shift that distant assembly line into high gear. You do that sort of thing in the precision business and things really go crazy. Since I'm already married to a lovely woman, I said, "It's a waiting game." He didn't want to hear that. He wanted me, Mr. Know-It-All, to tell him how to get his telescope within a week.

All I could do was sigh, and remember my hard times.

If quite a few telescope manufacturers share a major fault, this is it: They fail to cut themselves enough slack when stating delivery times. I'm sure the companies do not do this intentionally. Certainly, it is frustrating to be kept waiting on the buyer's side — but it is also frustrating for those on the seller's side who must sweat out deadlines, employee absenteeism and negligence, and 22,315 other things that can and do go wrong with maddening regularity. Murphy picks on the manufacturers, too.

Granted, quite a few telescope manufacturers tend to bite off more orders than they can chew — but they are being force-fed. What's needed is a little more sensitivity and understanding on both sides. The seller must be a bit more up-front and realistic while making his sales pitch, and the buyer must learn to control his or her "I Want It Now!" appetite.

Before you judge a company for its tardiness, put yourself in their place. Walk in their moccasins. For a few moments, you be the Sales Manager of fictional Precision-Eye Optics, okay?

"Okay."

You know it is possible, although not probable, that your company can get a telescope fabricated for Mr. Beginning Astronomer in four months. You must believe that because your company has just printed ten thousand expensive color catalogs stating that in bold print. However, in the last two months, there has been a veritable avalanche of orders. Word of mouth has it that Precision-Eye offers one of the best priced, highest quality observing packages in the business, and you know that's true. Also, your ad agency has done one heck of a good job; their campaign has literally given Precision-Eye telescopes sex appeal. To deal with the unexpected surge of orders, your company has been forced to hire new help on the assembly line and badly needs two more experienced, qualified opticians to boot. Since no qualified opticians have come out of the woodwork (even though you've spent a good deal of time searching for them), the ones you do have are now working overtime. You know the new guys on the assembly line need time to get acclimated to this precision business, and sometimes the new help makes mistakes. You have this gut feeling that if there's the slightest hitch, delivery of a new model Precision-Eye will take six months, not four. If any of your suppliers — and there happen to be twenty-three of them — fail to deliver component parts or raw materials, that delivery date could stretch to over a year. The magic word here is "could."

Now consider: you've got this excited, gung-ho client on the phone who wishes to order a Precision-Eye telescope. He's got pen in hand ready to make out a check for payment in full. The first thing he wants to know is: Can you get it to me quickly?

Okay, Mr. Sales Manager. You know what the man wants to hear. You know that if everything goes smoothly, you will be able to get him his scope in four months. Do you turn pessimistic and tell him it might take over a year, and risk losing a sale? Or do you tell him that, if all goes well, it could take four months, and hope for the best? Remember now, this job puts food on your table and pays your daughter's dental bills. You're human. You tend to protect number one. Be honest. What would you say?

I've taken quite a bit of flak for speaking up for the telescope manufacturers concerning this problem. I sympathize with every enthusiast who has had or will have difficulties getting their orders filled promptly. If the manufacturers did not keep coming up with wonderful new telescopes and accessories, I'd say let's see some heads roll. But they do keep coming up with great stuff and the end is not in sight. Just take a look at what is available on today's market, and compare this with what was available in the 60s. It's a wonderful time to be an astronomy enthusiast.

I sincerely believe that the buyer — not the seller — should be willing to go the second mile as long as he or she receives a quality product at the end of that second mile. Patience is the word. Everyone agrees that you can't rush quality — but that doesn't seem to deter everyone from doing their darnedest to rush it.

If, after the second mile, you still do not get results, feel free to take more drastic actions. Be aware that if the manufacturer cannot make shipment within 30 days of the projected delivery time, you have the right to cancel the order and receive your money back in full. If difficulties arise, furnish documentation to the Federal Trade Commission's Office of Consumer Protection and to the magazines which carry the manufacturer's ads. Just don't be too quick to fly off the handle. There are two sides to every story.

Now, I want you to change your identity: you are now Mr. Impatient Beginner Astronomer. For you, four months is much too long to wait, so you've decided to order from a manufacturer who promises off-the-shelf delivery. But you should be aware that for that luxury, you either pay more money or you take a gamble. Some companies will argue about this, but the special attention in the former case, and the mass production involved in the latter case, dictate it. In the case of expensive state-of-the-art equipment, you pay a high premium for the special attention of quick delivery. In the case of a mass-produced telescope, you pay a competitive price, get the scope quickly, and take a chance, albeit slim, of getting a lemon.

Monetarily, I'm not a rich man. I took a gamble once, and purchased a telescope from a fast-growing company who promised off-the-shelf shipment, guaranteed precision drive, and the finest null-figured optics: in essence, the finest telescope available for the money. What I received, admittedly in very short order, was a very sick telescope. The drive would not budge because of internal corrosion, and the mirror suffered greatly from a turned-down edge. It was painfully obvious to me that the telescope had not been tested. And yes, it made me angry to pack the thing up, spring for the return postage, and send it back.

Would you like to know whose fault it was? The company had a bum working for them, someone who wasn't doing his job. He was the piece of grit that fouled up the clockwork. The company quickly refunded my money, apologized, and fired the bum.

It wasn't a pleasant experience for me or the company. Because I'm human (yes, that's true), I've never been able to order another telescope from them. Many of my astronomical friends own that firm's telescopes, and they are all very satisfied. But for me, the bitter aftertaste remains — I was burned and I still have the scars.

But that experience in no way diminished my desire to explore further into amateur astronomy. Instead, it made me more determined to succeed. It made me tougher.

Well, to get back to the angry man who couldn't get his telescope (remember him?). I urged him to be patient. I told him that I had dealt successfully with the company, that they made a great telescope and would not rush at the expense of quality. I assured him that I had never heard anyone complain about the company's product, and that he would be pleased in the long run.

That didn't help at all.

His anger would not be undone.

He canceled his order and purchased one of those horrid department store telescopes I warned you about earlier that absolutely could not live up to its claim of "1200x." And you can imagine what that did to his interest in astronomy. Gurgle, gurgle, down the tubes.

Another unworthy bit the dust.

And then, there was the retired man who claimed he couldn't see a blankety-blank thing through his 16-inch reflector. Well, I've never been able to afford a 16-inch reflector. What could I say to him?

Should I come right out and say, "It's your fault"?

That's not what these people want to hear. They want to be assured that, yes-oh-yes, it is the fault of the manufacturer, and the fault of everyone else who has painted a rosy picture of amateur astronomy and lured them into it all. They're bitter and they're mad and they want someone else to take the rap.

Well, I'm sorry folks, but I won't take it! And I don't think the telescope industry should have to take it, or my precious friends in the world-wide family of amateur astronomers. By golly, I know what's up there in the firmament. And I know what's out there in the form of excellent equipment designed to enable you to appreciate and study the heavens. And in each case, what's out there is fantastic! So, the only thing I can do is try to nip this anger in the bud — kill the rumor.

Listen up now if you think you'd like to pursue amateur astronomy any farther into the jungle. And those of you who have made the grade should listen up, too. You just might be able to nip some buds yourself. Pruning helps this family tree, for sure.

If you are worthy, the universe will not chew you up and spit you out. If you have the burning desire to surmount all obstacles in your pursuit of amateur astronomy, there is nothing anyone can do or say to stop you. If not, no one else can light your fire. Either your pilot light is out, or you never had one.

If the word "quit" is part of your vocabulary, then you should make tracks and get out of this jungle quickly. Forget astronomy.

If you don't care enough about amateur astronomy to spend a great deal of basic training doing your "book lernin," this hobby is not for you.

If you do not have enough common sense to start off gently with a telescope you can handle, seek ye pleasure elsewhere.

If you've never taken the time to learn the fundamentals of optical collimation, and don't intend to, then don't.

But don't blame the instrument.

Or its manufacturer.

Or "poor image quality."

Or the universe.

Don't blame anyone but yourself. And don't, for heaven's sake, begrudge others because they seem to enjoy what they're doing. They've paid their dues. They've approached this hobby, this science, in the proper way. They've toughed out basic training. They've sweated blood to get where they are, and they are now reaping the rewards. Every amateur astronomer is in some way battle-scarred, and they have the medals to prove it.

In amateur astronomy, there are privates and there are generals. I fancy myself a captain. One of these days, I will be a general, because I'm devoted to the cause 100%.

You say you want to be a general, too? You really, really do?

Great! Fine! Wonderful! But are you willing to work slowly, surely up the ranks, or is your heart set on becoming a ninety-day wonder? Ninety-day wonders have short survival spans in amateur astronomy.

You say you'd like to start out as, maybe, a major? You say you really don't want to mess with being any lower rank because you've got enough money to purchase high-ranking equipment?

Fine. Go ahead. Try it. Just don't blame me or the universe when everything crashes in on you.

Friend, I don't care how impressive you will look in your major's uniform with all its costume decorations. And the universe doesn't either. It is going to see through all that fluff and chew you up and spit you out. The very fact that you have a weighty 16-inch state-of-the-art reflector and five thousand dollars worth of accessories does not mean piddly-squat to the universe. If she's not obligated to the likes of Carl Sagan, she sure as shootin' is not going to feel obligated to you. You're like a marathon runner who has not yet run a good mile — the only thing you're going to get out of the foolish attempt is sore muscles and an aversion to running.

Amateur astronomy is not a race in the jungle; it is a slow crawl hindered by gravity, bad seeing, and unworthiness. There is no finish line. You must start out crawling uphill. There are no downhill grades until you've crawled a respectable distance. When and if you reach those downhill grades, it is almost a religious experience, a nirvana of sorts. But nobody — nobody — experiences that nirvana until they're built up thick calluses on their knees, steely-strong muscles, and a healthy respect for the course.

Wanna see the calluses on my knees? They're there. And I'm justly proud of them. No, maybe "proud" is not the proper word. Maybe I "appreciate" my calluses because they enable me to continue crawling on.

You see, you must approach the universe on your knees. She demands humility, respects perseverance, and abhors being taken for granted. She scares the Kool-Aid out of me sometimes — but because I love her so, she gives me a refreshing drink from time to time. Never does she allow me to get up and run, but she continues to let me crawl. That's all I need.

What sort of individual is content with crawling?

Any amateur worth his weight in optical glass is content with crawling.

Why?

He knows that crawling is forward progress. After all, the

amateur astronomer has no cause to rush. The difference between our crawling and our rushing is immeasurable when compared to the vastness of the universe. And too, an amateur astronomer seeks not to be a conqueror, but an understander. He or she is someone special who is willing to take the time and effort required to grow close to something infinite, awe-inspiring, and frightening. Amateur astronomers are painfully aware of their own weakness and frailties, and yet they celebrate their places in the scheme of things.

Above all, amateur astronomers are willing to crawl to the Creator's glory.

I will willingly and joyfully crawl until the day I die. And then, I have the utmost faith that the universe will express her gratitude for my devotion by unfolding her secrets before me.

Unlike the poet Walt Whitman, I do not celebrate myself. I celebrate what is up there. I celebrate every star, open and globular cluster, planetary and gaseous nebula. I celebrate every planet, cratered wasteland, cradle of life, and gas giant.

I celebrate the most wondrous things.

That vicious rumor is a lie. The universe is not a lady of the evening. She cannot be bought. She is not a harsh mistress. She must simply be approached with intelligence, common sense, sensitivity, reverence, and humility.

She's a queen, and should be treated as one.

15 Litera-Touring the Jungle

THERE ARE VARIOUS AND SUNDRY places to get your hands on astronomy books and publications in this jungle. The beasties steer clear of the libraries — but such is not the case where bookstores and mail-order outlets are concerned. Here the beasties are more interested in feasting on your money than on your sanity.

At first, I intended to include the traditional bibliography — but since nothing about this astronomy book is traditional in any sense of the word, I thought, "What the heck! Nobody really reads bibliographies. Why not trick them into reading mine by making it look like a normal chapter in my 'normal book.'"

You're gonna need some books. A few at first, more as time goes by. I'm going to suggest that you buy several *real* books. I'm also going to suggest that you *not* buy some entirely fictitious titles, and hope you use some common sense where these are concerned and read between my lines.

Before you purchase any books, do these two things:

1. Subscribe to ASTRONOMY magazine, if you haven't already. (And no, the folks at AstroMedia did not suggest this. I did. This is my book, remember?) Write to them at: P.O. Box 92788, Milwaukee, WI 53202.

2. Subscribe to *Sky & Telescope* magazine. (Advanced stuff, granted, but you shouldn't be without it.) Write to them at: 49 Bay State Rd., Cambridge, MA 02238-1290.

The most reputable mail-order astronomy book outlets advertise regularly in the pages of these magazines. For my money, Willmann-Bell Inc. (a real firm, this time) offers the fastest, most dependable service and the widest selection this side of the Dog Star. And no, I *don't* have a piece of their action. I just wish I did.

Some of the books I'm about to suggest may be available at your public library. I strongly suggest that you *buy* them, however. Astronomy books are not like Stephen King novels — it takes years to devour good astronomy books from cover to cover, not an evening by the fire. A personal astronomy reference library is as useful as one of your arms. It can be a much-needed extension of your brain. It can also keep you darn good company on those rainy nights.

In my own humble opinion, the best astronomy text to date is, without a doubt, George Abell's *Exploration of the Universe*. Now in its fourth edition, it is a recognized classic, chock-full of enough easily understood hard-core astronomy to chew on for years. I have personally chewed up two editions, and am now munching on the latest one. How I wish I could write like Abell! I do not exaggerate when I say that just about everything an amateur astronomer needs to know about astronomy is in this book. Forego buying that extra eyepiece to get this book. Give up meals to get this book. It is that helpful, that deliciously good!

Norton's Star Atlas is another must-have. Don't wait to finish supper to order this book. It contains an excellent map of the sky, a reference handbook whose brevity belies its content, and a list of interesting objects to observe. It has been reprinted more times than I've got fingers. Recently, it has been revised and rewritten. It is easily affordable and a joy to use.

The magazine subscriptions and these two books are all you should need during your first year's foray into the jungle.

If things go well during this first year, your next purchase should be John Mallas' and Evered Kreimer's *The Messier Album*. This neat little volume will introduce you to some of the finest and most easily located deep-sky objects. Mallas gives accurate descriptions of the objects as seen through his 4-inch Unitron refractor — descriptions that you as an amateur can relate to. Also included are excellent sketches rendered by Mallas which show what you can expect visually. Many fine astrophotos taken by Evered Kreimer with his 12½-inch reflector round out the book, making it a most impressive and useful package.

If you still hunger for more books along this line, you might as well join the Astronomy Book Club and write off for all the catalogs that the booksellers who advertise regularly in the magazines offer. These outlets are virtual gold mines. Once you get on their mailing lists, you will not have to worry about missing a significant new book in the field. This does not hold true with your public libraries or your local bookstores.

Take advantage of Astronomy Book Club's introductory offer. Often you can get major titles for literally pennies on the dollar. That's how I bagged the next must-have title on my list — *Burnham's Celestial Handbook, Vols. I, II, & III.* I got all three in the hardback editions for $5.43, including shipping and handling — a $60 value! One of the best investments I've ever made.

Burnham's Celestial Handbook is truly what its subtitle says it is: "An Observer's Guide to the Universe Beyond the Solar System." It comprehensively covers literally thousands of celestial objects outside the solar system that are within reach of amateur-size instruments. The descriptive notes cover just about everything you would want to know, and there are over 600 photographs to provide icing for the cake. I stand in awe of Burnham for pulling it off. When you consider the obvious work involved in writing them, you *know* that he didn't do it for the money. The books radiate his love for amateur astronomy. Buy them.

If your interests, after the first year, lean toward planetary work, be advised that there are many, many excellent books concerning the planets always available (especially so in the light of recent advances in knowledge thanks to the successful planet probes of recent years). It seems that books cannot be printed fast enough to keep up with all the new information. This so, I suggest that you rely on the magazines to keep you informed about all the juicy gossip concerning the planets, and purchase a general book concerning all of the planets. My suggestion is *The New Solar System*, edited by J. Kelly Beatty, Brian O'Leary, and Andrew Chaikin.

"I can't believe you didn't suggest the Sidgwick books for the first year."

I did that for a reason. John Sidgwick's *Amateur Astronomer's Handbook* and *Observational Astronomy for Amateurs* are classics packed with an unbelievable amount of priceless information. You need a year's experience using your telescope before you can use these books to their best advantage. By all means, buy them — but wait until you're ready. Believe me, they will be around till Doomsday, and maybe even after.

If you survive the first year, you will feel the need to step up to a star atlas that considers stars of fainter magnitude than does

Norton's. Your best move here is to purchase Wil Tirion's *Sky Atlas 2000.0*. This atlas depicts 43,000 stars to the 8th magnitude and 2500 color-coded deep-sky objects with celestial coordinates for the year 2000. A companion *Sky Catalogue 2000.0* is available, but it's "for mature audiences only." Unless your work is very advanced, you probably won't use it that much.

"What if I decide to specialize in the Moon?"

Most books on the Moon are all made out of tikky-takky and they all look just the same. Before you buy any, bone up on geology at your public library. To really appreciate the Moon, you must appreciate the forces involved in forming it. The best way to do this is to apply what is up there to what is down here on Earth. In other words, you must become familiar with and study analogs of lunar structures here on Earth. Read about Earth's vulcanism, weathering, mountain formation, river basins, rilles, faults, and rocks. With a firm foundation in Earth-based geology, you should be ready to tackle advanced books on the Moon which will do more than tell you over and over things you have read many times before.

Once you get your feet wet in the geology phase, I suggest you purchase the very expensive *Moon Morphology* by Peter Schultz. Granted, *Moon Morphology* is aimed directly at the professional. It will take some work on your part to read it, but if you do your geology homework, you will understand much of what Schultz's clear writing will divulge. Unlike the advanced mathematics required to understand many advanced works on astrophysics, advanced work concerning the morphology of the Moon involves using your common sense. Try it; you'll see what I mean. If you're planning on specializing in the Moon, *Moon Morphology* is worth the effort and the expense. The fine Lunar Orbiter photos alone are worth the price of the book. If the book is out-of-print, ask the folks at Willmann-Bell to search for a copy, and be willing to pay them a premium if they find one.

Next, for your aesthetic enjoyment, purchase Hans Vehrenberg's *Atlas of Deep-Sky Splendors* and Timothy Ferris' *Galaxies*. Add to these Bart Bok's *The Milky Way* and Shapley's *Galaxies*.

If you're interested in amateur telescope making (and only if), good books along this line include: Richard Berry's *Build Your Own Telescope* (for those who walk the aesthetic path), and Jean Texereau's

How to Make a Telescope (for those who would follow the scientific path).

The books I've suggested will get you on the right track without loading you down with too much luggage. The more you get involved, the more you will feel the urge to spend money on titles like: *A Survey of the Membership of the American Gastronomical Society Concerning the Mutual Effects of Increased Stomach Acidity and Sky Transparency When Connected to Long-Exposure Astrophotography in a Warm and Humid Environment.* These things are usually written in unknown tongues and printed on brittle newsprint. On the average, they run 2,345 pages and cost $187.50. They don't even make good doorstops because they tend to crumble after about three months.

Esoteric titles may lure you: *Jupiter and Its Effects on Black History* by Clive Wilcox, an amateur astronomer who once shook the hand of the Reverend Martin Luther King, Jr. Or Fritz Plutarch's *Atlas Presidentium Regurgitorium,* an atlas of the sky as it might have been had Wendell Wilkie been elected president. Or *The Moon as Depicted on North American Quilts* by Grandma Galileo, a direct descendant of you-know-who. Forget them all. Use your head, okay? Buy books that will help you to observe more profitably, not books that will impress your friends.

Authors, if I didn't include your book, please don't jump on my case. I didn't even include mine! I couldn't possibly include every title available. If I did, this book would have to be printed on brittle newsprint, run 2,345 pages, and cost $187.50!

Some books did not make my list because they are so blasted boring. Some did not contain even a spattering of what I thirst for when I read: a sense of wonder. Some did not make my list because I couldn't afford them. Some because I haven't read them. Some because I couldn't understand their titles (that's a strong indication that what's inside will be even more elusive).

If I've made any of you struggling or non-struggling authors out there mad by neglecting to mention your favorite child, feel free to do me the same disservice. What's good for the goose is good for the gander.

I can take it. I'm tough.

16 Some Final Thoughts

WE'RE DEEP IN THE JUNGLE NOW, and it's just about time for me to leave you . . . to let go of your hand. You should be tough enough now to safely continue on your own. Just remember the following:

Be patient and understanding with yourself. Let's face it, there are going to be some crystal-clear nights when you just won't feel like setting your telescope up, much as it wants to be set up. On nights of fine seeing, a nice telescope will cry to you from its storage closet and make you feel guilty for not taking advantage of the night.

More often than not, you will listen for a while, give in, and suddenly disappear from events like your little boy's first speaking part in the school play, or your first family reunion in twenty-six years, or your pastor's first visit since you joined the church thirty-three years ago, or the rerun of the fifth installment of *Cosmos*. Doing so can do several things, all of them potentially threatening. It can make your spouse, if you still have one, begin to have bad thoughts about your telescope (the "spouse scorned" sort of thing). It can make those close to you wonder who you are ("Mommy, didn't he used to hang around here a lot?").

Yes, you can easily become an astroholic if you are not tough with yourself. Remember that moderation is the word. Too much of a good thing, like too much carrot cake, can turn your stomach real fast. You must learn to be firm with your telescope. Firm! Even if it means walking right up to its storage closet (don't open the door) and shouting, "Not tonight! And that's final!" It's a harsh thing to do, I'll grant you, and it will earn you some stares of dismay, especially at family reunions, but it is a helpful, cleansing gesture. And its works. Telescopes respect authority.

Be patient and understanding with your relatives and friends. Let's face it again, if you observe from your backyard, you immediately feel less friendly toward your neighbor, especially if he leaves his backyard light on because his registered collie is afraid of the dark. During those years B.T. (Before Telescope), you didn't even know your neighbor had a backyard light. Now the thing is an obscene intrusion on your right to darkness. And you'll just know he's doing it to aggravate you. That's why it comes on right after you've spent the better part of an hour aligning your scope to the pole star.

Suddenly, you'll want to build a solid cedar fence twenty-seven feet high and find out that will cost as much as a dome, both of which you can't afford. You begin thinking barbaric thoughts. Must you turn into Genghis the Slayer to protect your hobby? Of course not!

You're a civilized amateur astronomer in a civilized jungle, so get tough with yourself, not others. Your neighbor just might turn off that light if you ask him kindly. And there are those special filters that will help you deal with vapor lights. And when all else fails, you can always load up the scope like thousands of other dedicated enthusiasts do and brave the wilds under dark skies. Sure, you have to get a little tough with yourself to go to this trouble, but who ever said amateur astronomy was easy?

And then you have to deal with your relatives. If you're not careful, you're gonna feel the same way about them. The only person in your family who cares anything about astronomy is your nephew Pete, who works for the Air Force in another state and only visits during monsoon rains. You'll get angry with Uncle Jack because he can't see a blamed thing in that 4mm eyepiece. Granny Schmidt is gonna get on your nerves because she doesn't want to go out in "that there cold" to see that red glow that just might be volcanic activity on the floor of the crater Plato. You're gonna bite your tongue when your clumsy cousin stubs his toe on your tripod, knocks your alignment out of whack, and spills his Dr. Pepper down your tube assembly. And, of course, your allergic father-in-law is bound to sneeze on your best Erfle eyepiece.

You're gonna get disgusted with everybody in the whole world who couldn't give a tad about an occultation, or Jupiter's darkening Red Spot, or the fact that the Moon just came up upside-down. You are going to suddenly realize, and live up to the awful revelation, that there are actually human beings, *live* human beings, out there who just do not care. There will be those who criticize you for wasting your money on such things while there are children starving in Zimbabistan.

Little cuts like these will take their toll, and if you are not tough with yourself, you just might start agreeing with everyone. There is no easy way out of this problem. You have to be gentle and tough with yourself at the same time. You'll have to suffer the bad and relish the good if you truly want to be an amateur astronomer. Just resolve not to expect so much from others and be content with your love affair as it is: a one-to-one relationship between you and the universe. The rest will work itself out. With time, you will find someone who appreciates astronomy as you do. Who knows? It may turn out to be Uncle Jack!

Amateur astronomer is not a degrading term. On the contrary, it is a badge of honor. A badge of merit. It means you can handle the jungle. It means you are a survivor. It means the universe has deemed you worthy.

Amateur astronomy does not promise you a rose garden. It does not promise fame or fortune. All it promises is a chance to travel from one end of the universe to the other in a single night: a marvelous, magical, sight-seeing adventure available to anyone who is not afraid to get tough with himself. The universe does not suffer a sense of obligation to anyone. Remember that. She is there. She is awesome. She allows herself to be observed and commands appreciation and respect from all who sincerely look upon her wonders. But she does not demand our attention; she does not need our worship.

For all those who strive to embrace her majesty, she will, with time, grant certain favors. If you are persistent and worthy, there will be times when she will smile down upon you and whisper special secrets which only you will understand.

That's her way of saying, "I appreciate your caring."

That, my friend, is when your toughness gains you star stuff!

***

You're on your own. Go gently into that good night. I'm off to do my own thing, to explore regions of the jungle I've never explored before, and to enjoy those sections familiar to me.

Perhaps, if you prove worthy and survive, we'll meet again.

Happy star trails, and God speed!